39.95

ATLAS OF NEPTUNE

This first atlas of Neptune presents a collection of unique images taken by Voyager 2 during its dramatic encounter with this, the most remote of the major planets of our solar system.

A lucid and stimulating text tells the fascinating history that led to the discovery of Neptune, before describing the Voyager mission in detail and the latest information that has come from it. In this way the reader can appreciate the dramatic leap in knowledge and understanding of Neptune that planetary scientists gained from this space-probe.

Many of the best pictures available from NASA are used here for the first time. These explore in unprecedented detail the upper atmosphere of this dynamic planet, its ghostly rings and the pitted surfaces of its remarkable satellites.

With no more space-probes to the outer planets likely for several decades, this atlas is of lasting value and will become a standard reference to Neptune for the enjoyment of both professionals and amateurs.

Atlas of
NEPTUNE

Garry E. Hunt and Patrick Moore

CAMBRIDGE
UNIVERSITY PRESS

Published by the Press Syndicate of the University of Cambridge
The Pitt Building, Trumpington Street, Cambridge CB2 1RP
40 West 20th Street, New York, NY 10011–4211, USA
10 Stamford Road, Oakleigh, Melbourne 3166, Australia

Cambridge University Press 1994

First published 1994

Printed in Great Britain at the University Press, Cambridge

A catalogue record for this book is available from the British Library

Library of Congress cataloguing in publication data

Hunt, Garry E.
    Atlas of Neptune/Garry E. Hunt and Patrick Moore.
        p.   cm.
    Includes index.
    ISBN 0–521–37478–2
    1. Neptune (Planet)–Atlases. 2. Voyager Program. I. Moore,
Patrick. II. Title.
QB691.H86 1994
523.4'81–dc20
93–1778 CIP

ISBN 0 521 37478 2 hardback

Contents

Preface

The end of the first stage in Man's exploration of the Solar System came in August 1989, when the American space-craft Voyager 2 flew past Neptune and sent back our first really reliable information about that remote world. No more probes to the outermost planets have been planned as yet, so that it seems timely to produce a book which summarizes our knowledge of Neptune as concisely as possible. We are unlikely to learn much more for several decades at least.

As on previous occasions, the book was planned by both of us, working together. Hunt (one of NASA's Principal Scientific Investigators for the planetary missions) has been primarily responsible for the sections dealing with Neptune's interior, meteorology, magnetosphere and rings, as well as the description of Voyager itself; Moore has dealt with the descriptive and historical chapters, and the satellites. We hope that the result will be of interest.

G.E.H.
P.M.

Picture acknowledgements

The publishers gratefully acknowledge the help of the following individuals and organisations for allowing us to use their material in this book. Every effort has been made to obtain permission to use copyright materials; the publishers apologise for any errors and omissions and would welcome these being brought to their attention.

27 Catalina Observatory; 38, 39, 40, 41, 42, 43, 44, 45, 46, 47, 49, 54, 55, 58 NASA/JPL; 58 McDonald Observatory; 59, 60, 62, 63, 64, 65, 66, 67, 68, 71, 72, 73, 74, 75, 80 NASA/JPL

Illustrations drawn by Paul Doherty

Introduction

The giant blue world Neptune is the outermost known member of the Sun's family of planets. It was also the last to be surveyed from close range by a space-craft: Voyager 2. The date was 25 August 1989; the time was 0650 GMT, or just before midnight Pacific Daylight Time at Pasadena in California, headquarters of the whole Voyager mission.

The Neptune fly-by marked the culmination of a programme which has been an outstanding success by any standards. Voyager 2, launched in 1977, had already passed by Jupiter in 1979, Saturn in 1981 and Uranus in 1986. Its performance had actually improved with age, and the pictures and information sent back from Neptune were better than anyone had dared to hope. Only a few years earlier, the chances of success at Neptune had been officially estimated as no better than 40 per cent.

For once in a way, Nature had been helpful. Toward the end of the 1970s the four giant planets had been arranged in a long curve which made it possible to send the same probe from one to the other, using the technique known to scientists as 'gravity assist' and, rather less reverently, as 'interplanetary snooker'. Voyager 1 had been content with surveys of Jupiter and Saturn, together with their satellite systems, but Voyager 2 added Uranus and Neptune as well, providing us with our only detailed knowledge of these two remote, bitterly cold giants.

Neither can we hope for much new information in the near future. The fortuitous line-up of planets will not recur for over 170 years, so that the gravity-assist technique cannot be used in so complete a way, and journeys to the outermost regions of the Solar System will take longer. Moreover, there is the problem of finance. NASA's budget has been systematically attacked by the politicians, and there have been many cancellations, notably the proposed probe to Halley's Comet. There have also been disappointments, of which the greatest has been the faulty performance of the Hubble Space Telescope. No more American space-craft to Uranus or Neptune have been proposed, and neither have the Russians given any hint that they mean to explore the outer planets. Missions to Venus, Mars, Jupiter and Saturn are planned, but that is as far as can be foreseen at the present moment.

Therefore, this seems to be the right time to summarize what Voyager 2 has told us about Neptune, so that this new book may be regarded as a sequel to our account of Voyager 2 at Uranus.[†] Both the writers were at Mission Control during the fly-by, and it was very clear that we were witnessing the end of an era – the period which began in 1962 with the first results from Venus, and which has been so remarkably fruitful. As with Uranus, we expected surprises, and we were not disappointed. Neptune proved to be a dynamic world, and with its ghostly rings and its remarkable satellites it provided a fitting conclusion to the first stage in our practical exploration of the planets. Perhaps the most apt comment was made by Dr Laurence Soderblom, one of the leading planetary geologists, as Voyager completed its final task and began its never-ending journey into deep space: 'Wow! What a way to leave the Solar System!'

[†] *Atlas of Uranus*: Garry Hunt and Patrick Moore, Cambridge University Press, 1989.

The Jet Propulsion Laboratory: Pasadena, California.

Neptune in the Solar System

Five planets – Mercury, Venus, Mars, Jupiter and Saturn – have been known since very ancient times. This is not surprising, because all are prominent naked-eye objects (even Mercury, when at its best), and their movements mark them out at once from the so-called 'fixed stars'. Together with the Sun and Moon, this made seven members of the Solar System, and since seven was the mystical number of the ancients it seemed reasonable to assume that the system was complete. It came as a distinct shock when, in 1781, a then-unknown amateur astronomer named William Herschel discovered a new planet moving far beyond the orbit of Saturn. Uranus – named in honour of the first ruler of Olympus – is just visible with the naked eye if you know exactly where to look for it, but it is not at all conspicuous. Though it had been seen on numerous occasions before Herschel identified it, it had always been mistaken for a star.

Slight irregularities in the movements of Uranus led to the tracking-down of another planet, Neptune, in 1846. Giant though it is, Neptune is so remote that it is below naked-eye visibility. Binoculars will show it, and it is an easy telescopic object, but its disk is so tiny that to the casual observer it appears star-like unless a high magnification is used on an adequate telescope. Neptune has the distinction of being the first planet to be 'discovered' before it was actually identified telescopically. The whole story is somewhat complicated, but at any rate it provided a final vindication of the principles laid down by Isaac Newton over a century and a half earlier.

Still the situation was not quite satisfactory. Minor irregularities remained, and new calculations were made, giving the theoretical position of yet another planet. In 1930 the new body, Pluto, was found not far from where it had been expected, but whether or not this was coincidental is by no means clear, because Pluto seems to be unworthy of the status of a true planet. It is small – its diameter is much less than that of our Moon – and it is a cosmical lightweight. Moreover, it has a strange orbit which is much more eccentric than those of the other planets, so that for part of its 248-year revolution period it is closer-in to the Sun than Neptune can ever be. This is the case at the present moment, so that between 1979 and 1999 Neptune, not Pluto, is the more remote of the two. However, there is no fear of a collision; Pluto's orbit is tilted at the unusually large angle of 17 degrees, and no close approaches to Neptune can take place.

Any casual glance at a plan of the Solar System shows that it is divided into two definite parts. First, there are four small worlds – Mercury, Venus, the Earth and Mars – which have various points in common; all are solid and rocky, and all are fairly dense. Their surface conditions are very dissimilar, as we

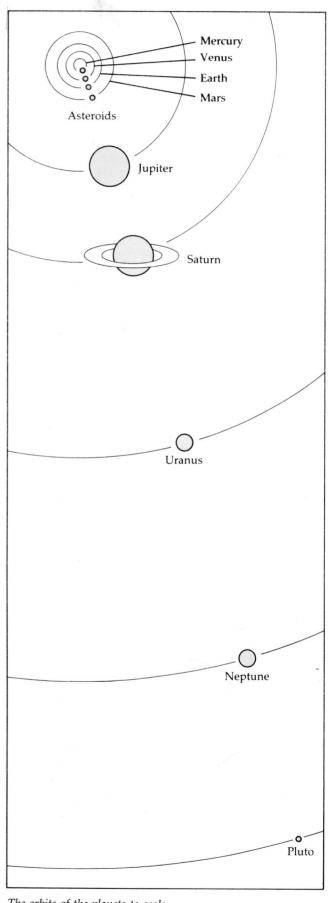

The orbits of the planets to scale.

know from the various space-craft which have been sent to them. Venus has a dense atmosphere made up chiefly of carbon dioxide, with clouds containing large quantities of sulphuric acid; the surface temperature is of the order of 500 degrees C, and life appears to be out of the question. The first survey was made by the American probe Mariner 2 in 1962, and since then there have been many others, of which the latest, Magellan, entered a closed path round the planet in August 1990. There are mountains, valleys, craters and (probably) active volcanoes; all in all, Venus is as hostile a world as can be imagined.

Mars is quite different. Here too there is an atmosphere made up mainly of carbon dioxide, but the density is low, with a ground pressure which is everywhere below 10 millibars. Mariner 4 sent back the first close-range pictures as long ago as 1965; in 1976 two Vikings made controlled landings, scooping up samples of Martian soil and sending back analyses. No traces of life were found, but the existence of dry riverbeds, together with huge volcanoes, indicates that there must have been a time when Mars was much more welcoming than it is now. Most authorities believe that the volcanoes are extinct; there are many craters and valleys, and the poles are coated with ice-caps, made up of a mixture of water ice and carbon dioxide ice. The temperature is, predictably, very low, but there is no reason to doubt that a manned expedition there will be possible fairly early in the twenty-first century.

From our point of view, the main disadvantage of Mars is its lack of breathable atmosphere; the planet is much less massive than the Earth, so that it has been unable to hold on to any dense atmosphere it may once have had. This is even more true of Mercury, whose atmosphere is negligible. Only one probe has been sent there: Mariner 10, which transmitted pictures in 1974 and 1975 showing a cratered, mountainous surface which looks superficially very like that of the Moon, though Mercury is close to the Sun and the daytime temperature is very high.

Beyond Mars, outermost of the inner planets, we come to the belt of asteroids or minor planets, only one of which (Ceres) is as much as 900 kilometres in diameter. Thousands of these dwarf worlds are known, but most are very small indeed, and may be regarded as little more than débris. It is not likely that they ever formed part of a larger body; much more probably no substantial planet could ever form in that part of the Solar System because of the disruptive effect of giant Jupiter.

Though most of the asteroids keep strictly to the main zone, between the paths of Mars and Jupiter, there are some which do not. Various small bodies swing inward, and may make close approaches to the Earth. Collisions are not impossible; indeed, there are serious suggestions that a major asteroid impact about 65,000,000 years ago caused such a change in our climate that the dinosaurs, which had been lords of the world for so long, were unable to adapt to the new conditions, and died out. Certainly we often encounter meteorites, which are irons, stones or a mixture – and there may well be no difference between a large meteorite and a small asteroid. One named asteroid, Hathor, can be no more than half a kilometre across.

There are also some asteroids which move well beyond the main group. The so-called Trojans share the orbit of Jupiter, though they keep prudently either well ahead of, or well behind, the giant planet and are in no danger of being engulfed. In 1990 a small asteroid was found in a similar 'Trojan orbit' with Mars. More significantly in our present context is the strange body known as Chiron – discovered by Charles Kowal in 1977 – which spends almost all its time between the orbits of Jupiter and Saturn. By asteroidal standards it is large, with a diameter of well over 100 kilometres, and in recent years it has shown definite signs of 'fuzziness', which presumably indicates a gaseous surround. Either Chiron is an exceptionally large comet or, more probably, it has a surface layer which is normally frozen, but which starts to evaporate when Chiron is at its closest to the Sun, producing a temporary atmosphere. Chiron's orbital period is 50 years; the date of its next perihelion passage (that is to say, its closest approach to the Sun) is 1996.

Chiron may be of special relevance here because there is a chance that it is the same type of body as Pluto, though Pluto, with its diameter of 2445 kilometres, is very much the larger. Moreover Pluto has a companion, Charon, which is comparable in size with Pluto itself, so that there is a temptation to regard the pair as a double asteroid. Neither can we rule out a past association between Pluto/Charon and the system of Neptune, for reasons to be discussed below. (Do not confuse Charon with Chiron. It is unfortunate that the two names are so alike.)

The four giant planets are widely spaced – a fact which is not always appreciated. All are rich in hydrogen, which is the most plentiful substance in the entire universe, but they are not alike, and in particular the Jupiter/Saturn pair is markedly different in composition from the Uranus/Neptune pair.

Telescopically, Jupiter shows a yellowish disk, crossed by dark streaks known as belts together with spots, wisps and festoons. The disk is obviously flattened, because Jupiter has very rapid axial rotation; the 'day' there is less than ten hours long. Since the surface is gaseous, Jupiter does not spin in the way that a rigid body would do. The rotational period of the equatorial zone is around five minutes shorter than that of the rest of the globe, and individual features have periods of their own, so that they tend to drift around in longitude. The surface details are always changing: Jupiter is in a state of constant turmoil, with violent winds and currents. The most famous discrete feature is the Great Red Spot, which the Voyager probes showed to be a vast whirling storm – a phenomenon of Jovian 'weather'. The cause of the strong reddish colour is not definitely known; it could well be due to phosphorus.

It is thought that Jupiter has a solid core at a temperature of the order of 30 000 degrees C, surrounded first by a deep layer of liquid metallic hydrogen and then by another deep layer of liquid molecular hydrogen, above which comes the 'atmosphere', made up chiefly of hydrogen but with a large amount of helium. There is a very powerful and complex magnetic field; the magnetic axis is inclined to the rotational axis by over 10 degrees. There are also very powerful radiation zones, and ever since the 1950s it has been known that the planet is a strong radio source. The ring system, unlike that of Saturn, is dark and obscure, and is not observable from Earth.

Jupiter sends out considerably more energy than it would do if it depended entirely upon the amount received from the Sun, confirming that it has a strong internal heat-source. Yet this falls far short of the amount needed to trigger off stellar-type nuclear reactions. Jupiter, massive though it is, is not nearly massive enough to be regarded as a potential star.

Saturn, like Jupiter, seems to have a silicate core, surrounded by liquid hydrogen above which comes the atmosphere, again chiefly hydrogen but with rather less helium than in the case of Jupiter. The rotation period is less than 11 hours, so that the globe, like Jupiter's, is obviously flattened; the overall density is actually less than that of water. Also like Jupiter, Saturn has a considerable store of internal heat, and the magnetic field is strong, but it is less powerful than that of Jupiter, and the magnetic and rotational axes coincide.

Saturn's surface features are of the same type as those of Jupiter, but are much more subdued, and there is nothing comparable with the Great Red Spot. Spectacular details are seen now and then, the best examples of modern times being a large white spot in the equatorial zone, seen in 1933 – discovered by the British amateur W. T. Hay (better remembered by many people as Will Hay, the stage and screen comedian) and another, found by the American amateur S. Wilber in 1990. Neither lasted for long, but both became prominent enough to be seen with small telescopes when at their best.

The glory of Saturn lies in its rings, which are made up of swarms of small, icy particles, all revolving round the planet in the manner of moonlets. The main system consists of two bright rings (A and B) separated by a gap known as Cassini's Division, in honour of the Italian astronomer who discovered it as long ago as 1675; closer-in to the planet is a semi-transparent ring, C (the Crêpe Ring) which has been known since 1850, and is not difficult to see with a modest telescope when the rings are tilted at a suitable angle to the Earth. However, the ring system is very thin, so that when placed edgewise-on to us it appears as a narrow line of light, too faint to be seen except with large telescopes and none too easy even then.

The Voyager images showed that the rings are very complex. There are indeed thousands of ringlets and minor divisions, and the brightest ring (B) was found to show strange radial 'spokes' which had certainly not been expected. Three extra rings were also detected outside the main system. Presumably the gravitational pulls of Saturn's satellites have a strong influence upon the overall ring structure, but many problems remain to be cleared up. Of special interest is ring F, beyond the main system. It is stabilized by two small satellites, Prometheus and Pandora – one just outside the ring, the other just closer-in – which keep the particles in their orbits, and act as 'shepherds'. In 1990, after further examination of the Voyager images, another small satellite, Pan, was even discovered within the Encke Division, a narrow gap in the outer bright ring (A).

It is generally said that the two outer giants, Uranus and Neptune, are twins. This is true insofar as size and mass are concerned, but there are important differences; in particular Neptune, like Jupiter and Saturn, has a strong internal heat-source, while Uranus has not. The strangest feature of Uranus is the tilt of the axis, which amounts to 98 degrees. This leads to a very curious 'calendar'; first one pole, then the other, has a 'night' lasting for 21 Earth years, with a corresponding 'midnight sun' at the opposite pole, though for the rest of the 84-year revolution period conditions are less extreme. The reason for this extraordinary inclination is unknown. It has been suggested that at an early stage in its career Uranus was struck by a massive body and literally knocked sideways. This does not sound very plausible, but it is difficult to think of anything better.

No doubt Uranus has a silicate core, but the internal composition is different from that of Jupiter or Saturn. Round the core it is thought that there is a dense layer in which gases are mixed with 'ices', i.e. substances which would be frozen at the low temperatures at the surface. These ices are mainly water, ammonia and methane, which condense in that order to form thick, icy cloud-layers. Methane freezes at the lowest temperature, and so forms the top layer, above which comes the predominantly hydrogen atmosphere.

When Voyager 2 approached Uranus, in 1986, it did so 'pole-on', so that the pole was in the centre of the apparent disk and the equator round the edge. There were no striking features, as with Jupiter or Saturn; all that could be seen were some inconspicuous clouds, though windspeeds were found to be very considerable. A major surprise concerned the magnetic field, which was fairly strong; the magnetic axis was found to be inclined to the rotational axis by 60 degrees, and to be markedly offset from the centre of the globe.

In 1977 a system of thin, dark rings had been discovered in a rather unexpected manner. Uranus passed in front of a star, and hid or occulted it; both before and after occultation the star was seen to 'wink', because it was being briefly hidden by the previously unsuspected ring system. Subsequently the rings were confirmed by Earth-based observations, mainly at infra-red wavelengths, and Voyager 2 gave good views of them. They are dark and narrow, containing a considerable quantity of 'dust'.

Neptune lay ahead. Before the Voyager 2 mission our knowledge of it was meagre, but it was assumed that there would be a magnetic field, together with radiation zones. The tilt of the axis was what we may regard as normal (28.7 degrees, as against 23.4 degrees for Earth), and the bluish colour was different from the obvious green of Uranus; in each case, however, the root cause was the absorption of red wavelengths by the atmospheric methane. No Neptunian ring system had been detected, though there were strong suspicions of incomplete rings or 'ring arcs'. Because of the marked internal heat, it seemed reasonable to assume that Neptune would have a surface much more active than that of Uranus, and the best Earth-based photographs did indeed show large patches which were presumably clouds of some kind.

All the giant planets were known to have satellite systems, but they were quite unlike each other. Jupiter had sixteen known attendants, of which four (Io, Europa, Ganymede and Callisto) are of planetary size; only Europa is smaller than our Moon, while Ganymede is actually larger than Mercury, though less massive. These four are known collectively as 'the Galileans', because the first systematic observations of them were made in 1610 by the great Italian pioneer Galileo, using his primitive 'optick tube' which was much less effective than modern binoculars, but which was good enough for him to make a series of spectacular discoveries – incidentally, confirming the theory that the Earth moves round the Sun and causing him a great deal of trouble with the established Church.

It has been said that 'there is no such thing as an uninteresting Galilean', and certainly the Voyager revelations were striking. Ganymede and Callisto proved to be icy and cratered, while Europa's surface is also icy but is as smooth as that of a snooker ball. Io, on the contrary, is violently active. On its red, sulphur-coated surface there are volcanoes which erupt all the time, sending 'plumes' high above the surface. Io also affects the radio emissions from Jupiter, and since it moves in the thick of the powerful radiation zones it must qualify as the most lethal world in the entire Solar System. All the other Jovian satellites are very small; the outer four have retrograde motion – that is to say, they move round Jupiter in a 'wrong-way' direction – and they are probably ex-asteroids which were captured by Jupiter in the remote past.

Saturn's satellite family is quite different. There is one large member, Titan, which is the largest satellite in the Solar System apart from Ganymede, and has a diameter greater than that of Mercury. It has a dense atmosphere which Voyager 1, the first of the 'twins' to by-pass Saturn, found to be made up chiefly of nitrogen – which, of course, makes up 78 per cent of the air we breathe. There is also a considerable quantity of methane, and altogether Titan is a remarkable world. As Voyager 1 approached it, there was a difference of opinion as to whether it could show the actual surface. One of the present authors (P.M) felt that we would have a clear view, while the other (G.E.H) maintained that Voyager would be unable to see through the clouds. The latter view proved to be correct, and we still do not know what Titan is really like. There may be oceans of liquid methane or ethane, with cliffs of solid methane and a methane rain dripping down all the time from the orange clouds above. We are not likely to find out until the year 2002, when a new space-craft, Huygens (part of the Cassini mission) is scheduled to make a controlled landing there.

Titan had a strong influence upon the planning of the Voyager missions. Studies of it were regarded as just as important as those of Saturn itself, but to make a close pass of Titan meant moving out of the ecliptic plane, so that it would not be possible to go on to Uranus and Neptune. This was the task of Voyager 1. If it had failed, then Voyager 2 would have been put into a Titan encounter path, and the outer giants would have been abandoned. It was therefore a great relief when Voyager 1 succeeded, leaving its twin to continue the journey to the boundaries of the Solar System.

Though Titan is the only large member of Saturn's family, there are other satellites between 300 and 1550 kilometres in diameter: Rhea, Iapetus, Dione, Tethys, Enceladus, Hyperion and Mimas. All are icy and cratered, though Enceladus has a surface which is curiously smooth in places, and Iapetus has one hemisphere which is bright and the other which is dark. The other satellites are much smaller; the outermost (Phœbe) has retrograde motion, and is presumably asteroidal. According to calculation, the peculiar body Chiron – which, as we have noted, moves mainly between the orbits of Saturn and Uranus – approached Saturn to within 16 000 000 kilometres in the year BC 1664. This is only 3 000 000 kilometres further than the mean distance between Saturn and Phœbe. Can this be significant? We have to admit that we do not know, but certainly Phœbe, with an estimated diameter of little more than 200 kilometres, is comparable in size with Chiron.

Uranus has no really large satellite. Before the Voyager 2 pass, five attendants were known: Ariel, Umbriel, Titania and Oberon, all between 1150 and 1600 kilometres in diameter, and the rather smaller Miranda. All were imaged by Voyager. As expected, the four larger satellites had icy surfaces with numerous craters, but Miranda proved to be incredibly varied, with plains, ice-cliffs and strange, roughly rectangular areas which have set geologists a series of problems. Ten new satellites were found, all close-in to Uranus and all small; two of them, Cordelia and Ophelia, act as 'shepherds' to the outer ring.

Pre-Voyager, two satellites of Neptune had been found, one large (Triton) and the other small (Nereid). Triton, unique among large satellites in having retrograde motion, was believed to have an appreciable atmosphere; Nereid had been shown to have a strange, eccentric orbit more like that of a comet than a normal satellite.

Such was the state of our knowledge when Voyager 2, having left Uranus far behind, approached its final target. But before discussing the latest results, let us look back in history to the strange and at times embarrassing story of how Neptune was discovered.

The discovery of Neptune

Who was the greatest observational astronomer of all time? One obvious candidate must be 'Friedrich Wilhelm Herschel', always known as William Herschel because in his young days he left his native Hanover to spend the rest of his life in England, and all his scientific papers were written in the English language.

Herschel was born in 1738. He came of a musical family, and trained as a musician; after a brief spell in the Hanoverian Army he travelled to London, and from there went to take up his first musical appointment, as an organist in Halifax. By 1766 he had moved to the fashionable resort of Bath, becoming an oboeist in the orchestra which played daily in the Pump Room. His musical career blossomed, both as a performer and as a teacher; within a few years he had become organist at the famous Octagon Chapel, and had also been made Director of Public Concerts in Bath. His sister Caroline joined him from Hanover, and life proceeded smoothly and satisfactorily.

Characters in the story of Neptune's discovery.

W. Herschel

But astronomy was coming to the fore. William's first observations were made as early as 1766, but it was not until 1772, when he had completed his first telescope, that the direction of his life really changed.

He concentrated upon making reflectors, and it was with one of these – a modest instrument, with a focal length of 7 feet (2.13 metres) – that he made the discovery which brought him instant fame. On 13 March 1781, from the garden of his home at No. 19 New King Street, Bath,[†] he was observing stars in the constellation of Gemini when he came across an object which was definitely not a star. It showed a disk, and he found that it moved slowly against the starry background from one night to the next. He believed it to be a comet, but before long it was found to be a new planet, moving far beyond the orbit of Saturn. After some discussion it was named Uranus – though it is worth noting that the Finnish astronomer Anders Lexell, who had been one of the first to show that the object was a planet rather than a comet, suggested calling it Neptune.

The orbit was soon worked out, and by 1802 the revolution period was given as 84.02 years – and since the real value is 84.01 years, this was to all intents and purposes correct. Yet even before then it had become obvious that something was wrong. Uranus was straying from its predicted course; there was a very considerable error.

The orbit had been calculated by using both the observations made since 1781, and also the older records when Uranus had been seen and had been mistaken for a star. (John Flamsteed, the first Astronomer Royal, had even given it a stellar designation – 34 Tauri.) In 1820 the French mathematician Alexis Bouvard computed a revised orbit, rejecting all the pre-discovery observations as being of dubious accuracy. Even this would not do. By 1832 Uranus was out of position by a full half-minute of arc, which was totally unacceptable. It was around this time that the first suggestions were made of a possible new planet which was pulling Uranus out of position.[‡]

One of these suggestions came from an amateur, the Rev. T. J. Hussey, Rector of Hayes in Kent, who went so far as to write to one of Britain's leading astronomers, George Biddell Airy. If a new planet existed, could its position be worked out from the perturbations of Uranus, and in that case could it be

---

† 19 New King Street is the only surviving Herschel house. It has been acquired by the William Herschel Society and turned into a small museum, which is open to the public and is well worth visiting.

‡ For a full historical account of the discovery of Neptune, see *The Planet Neptune*, by Patrick Moore: Ellis Horwood Ltd, Chichester 1989.

Sir G. B. Airy

J. C. Adams

optically identified? Airy's reply was not encouraging. Even if there were an unknown source of perturbation, he wrote, 'theory is not yet in such a state as to give the smallest hope of making out the nature of any external action' on Uranus. A few years later he was equally unenthusiastic in a letter to Alexis Bouvard's nephew, Eugène: 'If an unseen body were responsible, it will be nearly impossible ever to find out its place.'

Airy played a major part in the story of what has often been regarded as an unfortunate episode in scientific history. As Sir George Airy he was Astronomer Royal and Director of the Royal Greenwich Observatory between 1835 and 1881, and without question he was a great organizer; the Observatory had been beset with problems (largely because Airy's predecessor, John Pond, had been in poor health and had eventually been asked to resign), and Airy restored its reputation. He was also an expert instrument designer, and during his régime the Observatory was completely re-equipped. Yet Airy had his weaknesses too. He was obsessed with order and method, and it is said that he once spent a whole day in the Greenwich cellars labelling empty boxes 'Empty'. It was also claimed that to him, the action of wiping a pen on a piece of paper was as important as the setting-up of a new telescope. Of course this is a wild exaggeration, but there seems no reason to doubt the tale of how he used to insist upon his observers remaining on duty even when the sky was

hopelessly cloudy. Even when rain was falling he used to walk round the grounds after dark, visiting the various domes and saying to the observers 'You are there, aren't you?'

Neither was Airy disposed to change his mind once he had formed an opinion, even when it had become fairly clear that he was in the wrong. And this was the root cause of the whole trouble associated with the discovery of Neptune.

The Uranus problem would not go away, and in 1841 it was taken up by a young Cambridge undergraduate, John Couch Adams. On 3 July of that year he made a significant entry in his notebook:

'Formed a design, at the beginning of this week, of investigating as soon as possible after taking my degree, the irregularities in the motion of Uranus, which are as yet unaccounted for; in order to find whether they may be attributed to the action of an undiscovered planet beyond it; and if possible thence to determine the elements of its orbit etc. approximately, which wd probably lead to its discovery.'

He did pass his degree (brilliantly), and from that time onward Uranus was very much in his thoughts. By October 1843 he had completed most of his preliminary research, and had become quite convinced that he was on the right track.

Undoubtedly he was influenced by a mathematical relationship usually known as Bode's Law, though in fact it had been first noted by a less distinguished

Table 1. *Planetary distances*

Planet	Bode distance	Real distance
Mercury	4	3.9
Venus	7	7.2
Earth	10	10
Mars	16	15.2
(Ceres)	28	27.7
Jupiter	52	52.0
Saturn	100	95.4
Uranus	196	191.8

J. Challis

astronomer named Titius, and Johann Elert Bode, Director of the Berlin Observatory, merely publicized it. It may be summed up as follows:

Take the numbers 0, 3, 6, 12, 24, 48, 96, 192 and 384, each of which (apart from the first two) is double its predecessor. Add 4 to each. Taking the Earth's distance from the Sun as 10, the distances of the other planets, out as far as Uranus, are represented with fair accuracy, as shown in Table 1.

The Bode number 28 was represented by Ceres, the largest member of the asteroid swarm, which was discovered in 1801 and fitted neatly into the scheme. Therefore it was reasonable to assume that the proposed planet beyond Uranus would have a Bode distance of around 388. In fact, it does not; on the Bode scale, the distance of Neptune is only 300.7, so that the Law breaks down completely. In all probability the whole relationship is nothing more than coincidence, but Adams took it very seriously indeed.

His first step was to write to James Challis, Professor of Astronomy at Cambridge, who was helpful, and promptly wrote to Airy, saying that his 'young friend' Adams was working on the problem and needed some extra information. Airy sent the required data two days later, and Challis passed them on to Adams. In view of what happened later, Airy's initial promptness was rather ironical.

By mid-1845 Adams had obtained an approximate position for the new planet, and again Challis wrote to Airy, saying that the work was certainly of value, so that a personal meeting would be a good idea. Adams made two calls at Greenwich, but with no success. On the first occasion Airy was abroad, and on the second he was at dinner and could not be disturbed (in fact the time was 3.30 in the afternoon but Airy always dined at that time, and his rigid schedule was never altered). It seems that Airy was not even told that Adams had called. Adams did not try again, but sent a letter in which he gave the results of his calculations. Obviously he expected that a search would be put in hand.

U. J. J. Le Verrier

This, unfortunately, was not what happened. Instead, Airy wrote back asking a question which indicated – to Adams, at least – that he had not really appreciated the problem. Adams made no reply, and for some time nothing more was done.

Meanwhile, unknown to Adams, there had been developments from across the Channel. The problem of Uranus had been taken up quite independently by Urbain Jean Joseph Le Verrier, at the instigation of François Arago, Director of the Paris Observatory. Le Verrier was of a different temperament from the mild Adams; it has been said that he was one of the rudest men who has ever lived, though it may well be that history has treated him unkindly. Of his mathematical brilliance there was no doubt at all.

Two telescopes: Top *The Northumberland refractor, which did not discover Neptune;* Bottom *The Fraunhofer refractor at Berlin, which did.*

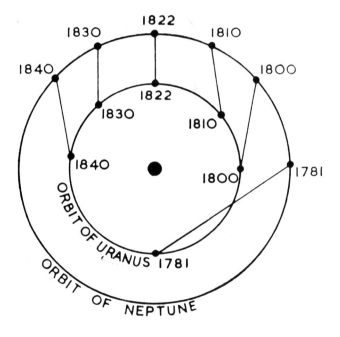

Relative positions of Uranus and Neptune, 1781–1840. From Uranus, Neptune was at opposition in 1822. Before that date, Neptune tended to accelerate Uranus; after that date, to retard it.

Le Verrier's method was rather different from that of Adams, but the final result was almost exactly the same. In 1845 he published a preliminary memoir, and in December a copy of this memoir reached Airy. In June 1846 Airy received a second memoir, and realized that the position given for the new planet was virtually identical with that given by Adams.

Airy wrote to Le Verrier, acknowledging the memoir and saying that

'I cannot sufficiently express the feeling of delight and satisfaction which I received from it'.

Yet he made no mention of Adams, and neither did he contact Adams personally. It seems that they passed by each other in early July when they were walking across St. John's Bridge in Cambridge, but their conversation lasted no more than a couple of minutes, and the Uranus problem was not mentioned at all, which is truly remarkable.

Yet Airy realized that the time had come to take action. There was no telescope at Greenwich suitable for carrying out a search (quite apart from the fact that any such programme would interrupt the routine work, a prospect which Airy viewed with extreme disfavour). So Airy wrote to Challis at Cambridge, instructing him to begin a survey with the 29.8-centimetre Northumberland refractor there. Again there were delays; Challis was abroad, and did not return until 18 July, so that his first telescopic sweep was not made until 29 July. He had no detailed star-map of the area in which the planet was expected to be, and he adopted the clumsy method of using a magnifying power of × 166, giving a field of view of 9 minutes of arc, and checking on all the stars which came successively into view. Adams had expected the planet to be at least as bright as magnitude 9. Actually it is about 7.7, but Challis was pessimistic, and decided to map all the stars down to magnitude 11. Before long he wrote to Airy:

'I get over the ground very slowly ... I find that to scrutinize thoroughly, in this way, the proposed portion of the heavens will require many more observations than I can take this year.'

Unknown to him, he was no longer the only searcher. Le Verrier had sent his results to the Paris Observatory, but no immediate action was taken, and by September his patience – never his strongest virtue! – had snapped. He sent his results to Johann Gottfried Galle, at Berlin, asking that a search should be made in the position indicated.

Galle made haste to consult the Director of the Berlin Observatory, Johann Encke. During their talk they were joined by a young student, Heinrich D'Arrest, who was fired with enthusiasm and asked to be

J. Galle

J. F. Encke

H. L. D'Arrest

allowed to join in. Galle agreed, feeling that it would be 'unkind to refuse the wish of this zealous young astronomer', and on the same night they began work, using the fine 23-centimetre Fraunhofer refractor which was possibly the best telescope of its kind.

Galle was no Challis. He had instinctive faith in Le Verrier's calculations, and he began by turning the telescope to the position which he had been given: Right Ascension 22 h 46 m, declination −13°4'. He hoped to find an object which showed a disk. Meanwhile D'Arrest had located the latest star-map of the area, and they settled down, with Galle at the telescope, calling out the positions and magnitudes of the stars which came into view, and D'Arrest checking them against the map. They did not have to wait for long. Within minutes Galle described an 8th-magnitude star at R.A. 22 h 43 m 25 s.84, and D'Arrest called out: 'That star is not on the map!' The hunt was over.

Encke joined them, and they tracked the object until it set. It did indeed seem to show a small disk, and by the following night they were sure; moreover, the object had shifted by just the expected amount. On 25 September Galle wrote to Le Verrier:

'The planet whose position you have pointed out actually exists.'

Encke followed this up with a letter:

'Allow me, Sir, to congratulate you most sincerely on the brilliant discovery with which you have enriched astronomy. Your name will be forever linked with the most outstanding conceivable proof of the validity of universal gravitation, and I believe that these few words sum up all that the ambition of a scientist can wish for. It would be superfluous to add anything more.'

Challis, still plodding on at Cambridge, heard the news of the discovery on 1 October. Only then did he start to check the observations which he had been making since the end of July – and realized that he had recorded the planet during the first four nights of observing. He had seen it again on 12 August, and if only he had compared his observations he could not have failed to make the discovery. On 23 September he had recorded it again, and had even suspected it of showing a disk, but had not taken the elementary step of putting in a higher-power eyepiece to see whether or not this was genuine. All in all, it must be admitted that he had been guilty of almost incredible incompetence.

The damage was done; the planet had been found, and the glory was Le Verrier's. But the political storm was about to break. On 3 October Sir John Herschel,

W. Lassell

F. W. Bessel

son of Sir William, wrote a letter which was published in the London *Athenæum*, pointing out that although Le Verrier's triumph could not be disputed, Adams had produced the same result considerably earlier. When the French realized that Le Verrier's claim to absolute priority was being disputed, they were furious. Arago at once published a thunderous onslaught:

'The friends of science will not permit the perpetration of such a flagrant injustice!... Mr. Adams has no right to figure in the history of the planet, neither by a detailed citation, nor by the slightest allusion.'

Further attacks followed, and for a time the whole affair threatened to blow up into a full-scale international incident. Fortunately neither Adams nor Le Verrier took any part, and when they first met face to face, in 1847, they struck up an immediate friendship which lasted for the rest of their lives, even though Adams could not speak French and Le Verrier was equally unversed in English!

There was also some controversy about a name for the new planet. Various suggestions were made: 'Neptune' seems to have come initially from Le Verrier himself. There was a short period when the name 'Le Verrier' was used, mainly in France, but the mythological system prevailed in the end, as it was bound to do.

There is an often-told story that William Lassell, a noted English amateur who had built himself a large and powerful telescope, was invited by Adams to take part in the search, but was unable to do so because he had sprained an ankle and could not go to his telescope. Actually, recent studies have shown that this story is almost certainly untrue. Yet it is strange that neither Adams nor Le Verrier undertook a personal hunt. They would have had no need of a large telescope, as one of the present authors (P.M) demonstrated not long ago. With the same magnification on the same telescope that Challis had used, Neptune's disk was unmistakable; moreover, a few nights checking with ordinary binoculars was sufficient to show the slow but definite motion of the planet against the background stars.

Today Adams and Le Verrier are recognized as the co-discoverers of Neptune. This is certainly fair, but it is also fair to add that the two astronomers who first identified it, and proved that it was a new member of the Sun's family, were Johann Galle and Heinrich D'Arrest.

Pre-discovery observations

Neptune is not a faint object – it is less than two magnitudes below naked-eye visibility – so that as soon as it had been identified, suggestions were made that, like Uranus, it might have been seen earlier and mistaken for a star. Obviously it was important to track down any such observations, because they would help in the work of computing an orbit. Remember, Neptune is a slow mover; it takes almost 165 years to complete one journey round the Sun, so that it was discovered less than one Neptunian year ago.

Apparently the first astronomer to check for pre-discovery observations was S. C. Walker, of the United States Naval Observatory, who had planned to search for Neptune but had been unable to get started before he heard the news of the success at Berlin. Walker examined old star catalogues, and found that the French astronomer J. J. de Lalande had seen Neptune on 8 May 1795 and again on 10 May. Since the two observations of what looked like an 8th-magnitude star differed in position, because Neptune had moved perceptibly during the two-day interval, Lalande had rejected the first as being faulty. In fact, he had Neptune within his grasp.

The planet also escaped John Lamont, often known as Johann von Lamont because he spent much of his career in Germany as Director of the Munich Observatory. He recorded Neptune three times: on 25 October 1845, and 7 and 11 September 1846. If he had compared the two latter observations he would have realized that he had found something of tremendous importance. Unfortunately for his reputation, this is precisely what he did not do.

Yet in our own time it has been established that there is one pre-discovery observation which goes back to the very first period of telescopic astronomy. It was made in 1612 by no less a person than Galileo!

Galileo, of course, was the greatest of all early users of the telescope, and from January 1610 he made a series of spectacular discoveries, including the mountains and craters of the Moon, the phases of Venus, and the myriad stars in the Milky Way. Of special importance were the four satellites of Jupiter, now called Io, Europa, Ganymede and Callisto – though these names were not widely used until modern times, probably because they had been proposed by Simon Marius, who claimed to have seen the satellites earlier than Galileo (and may well have done so). A few nights' work showed Galileo that the satellites were orbiting Jupiter, and this proved that, contrary to the old theories, there must be more than one centre of motion in the Solar System.

On 27 December 1612, when he was making a sketch of the positions of the Jovian satellites, he also recorded a 'star' which seems certainly to have been Neptune. An even more positive observation was

RECENS HABITAE. 27
min. 2. fec. 30. ab hac occidentalior diftabat min. 1. Vici̧

Ori.   *   O   * *   Occ.
   * fixa

niores Ioui exiguæ apparebant, præfertim Orientalis, extremæ verò erant admodum confpicuæ in primis verò occidua, rectámque lineam fecundum Eclypticæ ductum defignabant ad vnguem. Horum Planetarum pro greffus verfus ortum ex collatione ad prædictam fixam manifeftè cernebatur, ipfi enim Iuppiter cum adftantib. Planetis vicinior erat, vt in appofita figura videre licet. Sed Ho. 5. Stella orientalis Ioui proxima aberat ab eo min. 1.

Die 28. Ho. 1. duæ tantum Stellæ videbantur; orientalis diftans à Ioue min. 9. Occidentalis verò m 2. Erant

Ori.   *   O   *   Occ.
   * fixa

fatis confpicuæ, & in eadem recta: ad quam lineam fixa perpendiculariter incidebat in Planetam orientalé, veluti in figura. Sed hora 5. tertia Stellula ex oriente di-

Ori.   *   * O *   Occ.

ftans à Ioue m. 2. confpecta eft in eiufmodi côftitutione. Die 1. Martij Ho. 0. m. 40. quatuor Stellæ orientales omnes

Galileo's observations of Neptune, from Sidereus Nuncius. *There seems no doubt that the 'moving star' was, in fact, Neptune.*

made on 28 January 1614. A 7th-magnitude star was shown in the sketch, which has been identified with the star now catalogued as SAO 119234 and was marked *fix a*. Not far from it was another object, which must have been Neptune. Galileo's note reads, in translation: 'Beyond fixed star *a*, another followed in the same straight line, which was also observed on the previous night, but they then seemed further apart from one another.' Galileo had detected Neptune's motion. The telescope used seems to have had a magnification of ×18 and a field of view 17 arc-minutes in diameter. The actual magnitude of Neptune was 7.7, and Galileo often plotted stars fainter than that.

Because Neptune moves so slowly, at a mean rate of no more than 22 arc-seconds per day, it is not surprising that Galileo failed to realize the significance of his observation; moreover at that time the planet was in Virgo, a rather barren region of the sky, so that there were no bright stars nearby to act as comparisons. But the possibilities were there, and it is fascinating to think that Neptune could have been discovered almost one hundred and seventy years before the much closer and brighter Uranus. Fate can play some strange tricks.

Early theories of Neptune

Mention has already been made of William Lassell, one of the most distinguished amateur astronomers of the 19th century. By 1844 he had constructed a 61-cm reflector, which for that time was very large indeed, and he was also an excellent observer. As soon as he heard of the discovery of Neptune he began making observations, and in particular he wanted to see whether he could find any satellites. On 1 October 1846 Sir John Herschel wrote to him asking him to begin a search 'with all possible expedition'. Lassell obliged, and on 2 October he discovered Triton, the only large member of the Neptunian system. By 10 October he was quite certain of its existence, and he also suspected something else – a ring.

It had already become clear that Neptune was a giant planet, strictly comparable in size with Uranus and with a diameter less than one-half that of Saturn. It did not obey Bode's Law, which proved to be an astronomical red herring (which is probably one reason why the original orbits worked out by Adams and Le Verrier were wrong; Adams gave the revolution period as 227 years, Le Verrier as 217 years). A ring round Uranus had been reported by William Herschel, using his 20-foot (6 metre) focus reflector in 1787; although nobody else ever saw it, and Herschel himself finally realized that it was non-existent, lingering doubts remained. (We now know that the illusion was caused by the 'front view' optical system of Herschel's telescope; there was no secondary mirror, and the main mirror was tilted to bring the light directly to focus. The Herschelian system sounds convenient, but it has many disadvantages, and is never used today.) Therefore there seemed to be a chance that Neptune might have a ring, and on 2 October 1846 Lassell believed that he had seen one. On the following night he recorded that he 'received the impression of a ring, not much open, and nearly at right angles to the parallel of daily motion'. On 10 November: 'The planet very like Saturn, as seen with a small telescope and low power, but much fainter.' Other observers with him also believed that the ring could be seen, and Lassell sent a report to the Royal Astronomical Society, which took it very seriously; after all, Lassell had an excellent reputation, and his telescope was one of the largest and best in the world at that time.

One of those present at the RAS meeting at which Lassell spoke was John Russell Hind, who had been making regular observations of Neptune with the 17.8-centimetre refractor at a private observatory in Regent's Park. Subsequently Hind commented that 'the existence of a ring appears as yet undecided, though most probable', and James Challis, using the Northumberland telescope at Cambridge on 12 January 1847, called the ring 'very apparent' with a power of ×215. Yet other observers failed to see any ring-like appearance, and within a year or two even Lassell was becoming dubious. He transferred his large telescope from England to the clearer climate of Malta, and on 15 December 1852 he made the decisive observation. The angle of the supposed ring changed in position when he rotated the tube of his telescope, so that, in his own words, 'Whatever may be the cause, it is more intimately related to the telescope than the object'.

That was the end of the Neptunian ring. Lassell never reported it again; neither did anybody else. Lassell came to the reluctant conclusion that slight flexing of the primary mirror of his telescope had been the cause of the illusion, and one has to assume that Hind and Challis were guilty of 'wishful thinking', since it is only too easy to 'see' what one half-expects to see (remember the canals of Mars!). Note, however, that as soon as Lassell realized that he had been mistaken, he was honest enough to say so.

As soon as Neptune had been found, efforts were made to measure its rotation period. With a planet such as Jupiter this is easy enough; all that has to be done is to check on the surface features as they are carried from one side of the disk to the other by virtue of the planet's spin. Even Saturn shows enough detail to give a reasonable value. But Uranus and Neptune are different; telescopic observations failed to show any positive features on their tiny disks. Edward Emerson Barnard, one of the best and keenest-eyed planetary workers of the last century, made a special effort to see whether he could find any markings on Neptune, but using the world's two greatest refractors (the Yerkes 101.6-centimetre and the Lick 91.4-centimetre) he failed completely. The only chance of determining the rotation period seemed to be to look for slight, regular variations in the planet's brightness which could be due to rotational effects, assuming that some parts of the globe were darker than others.

Early attempts gave periods which we now know to be much too short: in the region from 8 to 10 hours. The first study which could be treated with any confidence at all was not made until 1928, by J. H. Moore and D. H. Menzel, using a spectroscopic method.

Just as a telescope collects light, so a spectroscope splits it up. Stars (including the Sun) are self-luminous, and show characteristic spectra; there is a bright rainbow band with colours ranging from red at the long-wave end through to violet at the short-wave end, crossed by dark lines, each of which is the trademark of one particular substance. Since a planet shines by reflected sunlight, it will show what is essentially a solar spectrum, but superimposed on it will be dark lines due to the atmosphere of the planet itself.

With an approaching body, all the lines will be shifted over to the short-wave end of the spectrum, while with a receding body the shift will be to the red or long-wave end (this is the well-known Doppler effect). If the planet is rotating, one limb will be approaching us and the other receding, so that the Doppler shifts will be in opposite directions. This is how Moore and Menzel attacked the problem. There are many complications to be taken into account, but they felt able to give a period of 15.8 hours, with a possible error of no more than an hour. Since the true period is now known to be 16 hours 3 minutes, they were remarkably close to the mark.

More importantly, it was clear that Neptune does not share the extraordinary axial tilt of Uranus. The inclination to the perpendicular to the orbital plane is 29 degrees, not much more than that of Saturn (26.7 degrees). Again, early estimates were not very far wrong.

Measuring the diameter of Neptune proved to be surprisingly difficult, because the apparent diameter is only just over 2 seconds of arc – and when using a device such as a micrometer to measure so tiny a disk, a very slight error can lead to a very large error in the derived diameter. Certainly Uranus and Neptune were similar, though it was soon found that Neptune is the more massive of the two (17 times as massive as the Earth, as against only 14 times for Uranus). However, all the diameter estimates clustered round 50 000 kilometres, with appreciable polar flattening due to the relatively quick rotation. It was only in our own time that a really precise value has been obtained. Neptune's equatorial diameter is 50 538 kilometres, and its polar diameter 49 600 kilometres. The equatorial diameter of Uranus is 51 118 kilometres, so that in size, at least, the two outer giants really are almost identical twins.

Neptune before Voyager

Because Neptune is so remote, our knowledge of it before 1989 was decidedly meagre. Voyager 2 told us more in a few weeks than we had been able to find out during the whole course of history. However, we had at least been able to establish some facts, quite apart from the planet's orbit, size, mass and rotation period.

Until well into the 20th century it had been widely supposed that all the giant planets, including Neptune, were miniature suns, giving off enough heat to warm their satellite systems. This intriguing idea was disproved in 1923 by a series of brilliant papers by Sir Harold Jeffreys, who showed that a more likely model was that of a cold gaseous layer above a solid surface. Another model was proposed in 1947 by Rupert Wildt, who assumed that each giant planet had a rocky core surrounded by a thick layer of ice above which came the gaseous atmosphere. Shortly afterwards came a theory due to W. R. Ramsey; this time Jupiter and Saturn were made up chiefly of hydrogen, with the lower layers showing the characteristics of a metal, while the less massive Uranus and Neptune had lost much of their original hydrogen and helium, so that they were composed largely of water, methane and ammonia. Much later – in 1980 – came a theory due to W. B. Hubbard and J. J. MacFarlane, which gave Neptune a silicate core 16 000 kilometres across, surrounded by a mantle of 'ices' made up of water, methane and ammonia, over which lay the low-density hydrogen-helium atmosphere.

One important difference between Uranus and Neptune was confirmed. The temperatures of the two are about the same: −214°C for Uranus, −220°C for Neptune. Yet Neptune is more than 1 600 000 kilometres further from the Sun, so that it receives much less solar radiation. This can only mean that Neptune has a strong internal heat-source, which Uranus lacks. Since Jupiter and Saturn also have marked internal heat-sources, it is Uranus which is the 'odd one out'. The reasons are unknown, though it has been suggested that the exceptional axial inclination of Uranus may have something to do with it.

Spectroscopic work showed that the main bulk of the outer atmosphere was hydrogen, which was no surprise; methane and ethane were also found. But it was not until 1979 that any cloud features were definitely recorded. Using one of the world's largest

Neptune 1991: Drawn by Patrick Moore, using a magnification of × 600 on the 60-inch reflector at Palomar. No markings are shown.

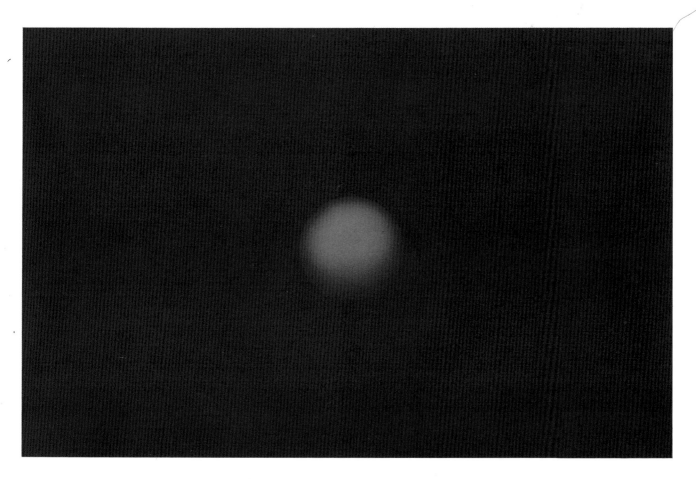

The Las Campanas Observatory: Top *General view;* Bottom *The dome of the Irénée du Pont Telescope.*

telescopes, the 2.5-metre reflector at the Las Campanas Observatory in Chile together with a modern electronic device known as a CCD (Charge Coupled Device) which is much more effective than a photographic plate, B. A. Smith, H. J. Reitsema and S. M. Larson identified clouds, and in 1983 they found that

'the Neptune images are characterized by several patches of haze located at the mid-latitudes in both northern and southern hemispheres ... Neptune appears to have a cloud structure'

much more marked than that of Uranus, which still appeared featureless, and whose clouds were not detected until the fly-by of Voyager 2 in 1986.

The Las Campanas Observatory: The telescope itself, used to take the best pre-Voyager pictures of Neptune.

Neptune, photographed with the 154-cm reflector at the Catalina Observatory (University of Arizona) by B. A. Smith, H. J. Reitsema and S. M. Larson.

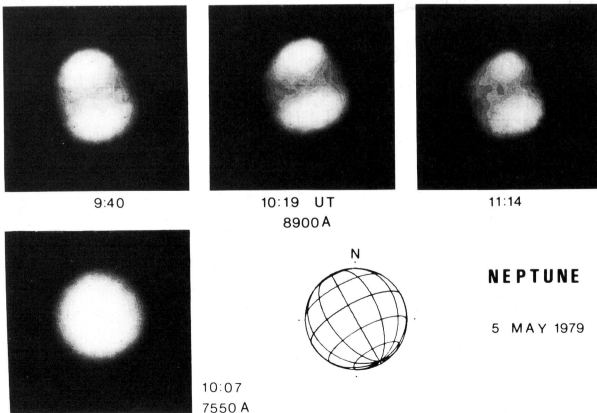

9:40   10:19 UT   11:14

8900 A

10:07

7550 A

N

NEPTUNE

5 MAY 1979

Because of the extra amount of internal heat, it was tacitly assumed that Neptune would be a more active world than Uranus, but nobody could be sure. It was also assumed that there would be a magnetic field, particularly after Voyager 2 had established that there is a strong field associated with Uranus. The discovery that the magnetic field of Uranus is inclined by 60 degrees to the axis of rotation was generally put down to some effect of the unique tilt of the rotational axis, and nothing similar was expected for Neptune.

Next, what about the possibility of a ring system?

The rings of Uranus had been detected in 1977 by the occultation method. Once found, they were confirmed both by subsequent occultations and by direct viewing. At the Siding Spring Observatory in Australia, D. A. Allen used the 3.9-metre Anglo–Australian Telescope at infra-red wavelengths, and produced striking views of the rings, but with Neptune he had no luck. 'The results were disappointing,' he wrote in 1983. 'The planet is too small to see clearly, and shows no hint of a ring.' It seemed, then, that the occultation method was the only hope.

Various inconclusive observations were made. During near-occultations of stars by Neptune some 'winks' were recorded, but not with any real certainty, and there was nothing similar to the effects seen for Uranus. It was suggested that there might be incomplete rings – 'ring arcs', in fact – though some astronomers were dubious; for example H. J. Reitsema commented that he was

'willing to accept part of a ring, but that creates real problems, because I don't understand how you get parts of rings'.

Two members of the team which had discovered the Uranian rings, J. L. Elliot and R. A. Kerr, went so far as to say in 1984 that

'if Neptune has rings, they will almost certainly not be discovered from the ground'.

We now know that rings exist, and that the so-called arcs are merely the brightest parts of them; one of the winks may have been due to the temporary hiding of the star concerned by one of Neptune's inner satellites, unknown before the Voyager 2 pass.

Finally, there was the question of the satellite system. Triton was easy to observe – easier than any of the satellites of Uranus – but its diameter was uncertain; estimates ranged between 6000 kilometres and only 2500 kilometres. It was thought possible that it might be considerably larger than the Moon, and to have a reasonably dense atmosphere, with a surface which could be partly or mainly covered with an ocean of liquid methane or even liquid nitrogen. In 1987 M. L. Delitsky and W. R. Thompson speculated that

'Perhaps Voyager 2, turning its cameras on Triton, will see plains of white and coloured organic deposits and, maybe, the glint of a distant sun reflected off a calm nitrogen sea'.

The truth proved to be very different, but in one way at least Triton was known to be unique among large satellites: it has retrograde motion. All the other retrograde satellites (the outer four in Jupiter's system, and Phœbe in Saturn's) are very small, and presumably ex-asteroids.

The other known satellite, Nereid, was discovered in 1949 by G. P. Kuiper on photographs taken with the 208-cm reflector at the McDonald Observatory in Texas. It had an estimated diameter of between 200 and 400 kilometres (the actual value is 169 kilometres), and had a highly eccentric orbit, with a revolution period of almost 360 days.

Such was the state of affairs when Voyager 2 drew in toward Neptune during the summer of 1989. All the equipment was functioning well, and the scientists assembled at the Jet Propulsion Laboratory at Pasadena expected some fascinating revelations. They were certainly not disappointed.

The Voyager mission

The Voyager missions to the outer planets will always rate among the greatest achievements of the space programme. They were not the first probes to visit the depths of the Solar System – Pioneers 10 and 11 had by-passed Jupiter earlier, and Pioneer 11 had also made a preliminary reconnaissance of Saturn – but during the 1970s and 1980s the two Voyagers provided what may be termed a quantum leap in our knowledge of the giant planets.

Voyagers 1 and 2 were identical twins, travelling in complementary paths through the outer Solar System. Both were launched from Cape Canaveral, in Florida. Voyager 2 went first, on 20 August 1977, and Voyager 1 followed on 5 September, though Voyager 1 was moving in the more economical orbit and reached its target first. It by-passed Jupiter on 5 March 1979, and Saturn on 12 November 1980. It is now on a path which has taken it well away from the plane of the ecliptic, so that it will have no more planetary encounters; however, the engineers at Mission Control are still in regular contact with it, and are receiving unique information from a region into which no space-craft has ventured before. All being well, contact with it should be maintained until it reaches the heliopause, i.e. the boundary of that part of the Galaxy in which the Sun's influence is dominant. Voyager 2 encountered Jupiter on 9 July 1979, Saturn on 25 August 1981 and Uranus on 24 January 1986 before its epic pass of Neptune on 25 August 1989.

Collectively, the Voyagers have produced a multitude of discoveries. Among them are new satellites of all the giant planets; the narrow, dusty Jovian ring; volcanoes on Io; amazing geological features on the various satellites, notably the extraordinary Miranda with its rich variety of terrain; colourful meteorological phenomena in the atmospheres of Jupiter and Saturn, with windspeeds reaching 1000 miles per hour on Saturn – that is to say, over 1600 kilometres per hour; evidence for an extensive nitrogen-rich atmosphere on Titan; a tilted off-centre magnetic field for Uranus – the list seems almost endless. Voyager 1's success at Titan was particularly important. Had this not been achieved, then Voyager 2 would have been programmed to survey Titan as well as Saturn, and would not have been able to go on to Uranus and Neptune.

Full-scale model of Voyager 2, at the Jet Propulsion Laboratory, Pasadena.

Space-craft and instruments

The Voyagers were based on the highly successful Mariner design used on so many earlier reconnaissance missions to the planets. Each space-craft weighs 825 kg (1819 lbs), carries 11 scientific instruments and is dominated by the huge 10-metre magnetometer boom. The on-board computer systems are capable of directing the scientific instruments and engineering equipment which control all the operations. Solar energy cannot be used as a source of power in these remote regions – there is simply not enough sunlight! – and so each Voyager carries its own source of electricity in the form of three Radioisotope Thermoelectric Generators (RTGs), which are miniature nuclear power-plants able to convert about 7000 watts of heat into 400 watts of electricity. At launch the power output from the RTGs was 475 watts, but obviously this could not be maintained. Plutonium oxide is used, and this decays at a steady rate; also there is degradation of the silicon–germanium thermocouples. Therefore the power output decreases at about 7 watts per year. Fortunately there is enough reserve to allow the Voy-

agers to function until at least the year 2017, while the propellant should last until 2035 – though whether contact will be maintained for as long as that seems rather dubious.

When in flight, each space-craft is three-axis stabilized, using the Sun and a bright star (usually Canopus) as celestial references. Behind the antenna is the main 'bus', a ten-sided aluminium framework containing the space-craft's electronics. This structure surrounds a spherical tank containing the hydrazine which is used for manœuvring the Voyager during flight. Although the space-craft is a very complex machine, it cannot think for itself; it must be told what to do, what sensor to use and so on. A set of instructions is generated in advance to tell the sensor what to do and when to do it.

Remember, too, that the Voyager is receding rapidly from the Earth all the time, so that things become progressively more difficult. At Neptune, the

A perspective view of the Voyager 1 and 2 trajectories from launch through the time of Voyager 2's encounter with Neptune.

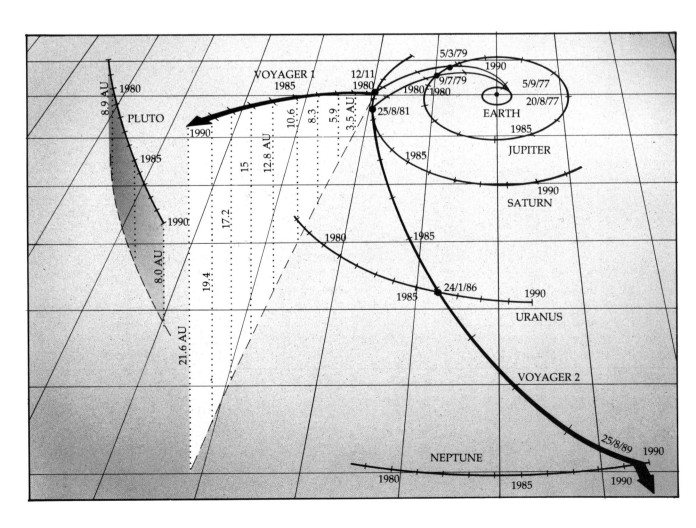

one-way communication time from Voyager 2 to Mission Control in Pasadena was 4 hours 6 minutes. Moreover, the amount of power received from the space-craft was almost incredibly small.

The three key sub-systems for controlling the engineering aspects of the mission, and supervising the scientific investigations, are the Computer Command Sub-system (CCS), the Flight Data Sub-system (FDS) and the Attitude and Articulation Control Sub-system (AACS). (Space technicians are notoriously fond of acronyms, and these may sometimes appear confusing – but one soon becomes accustomed to them, and it will be best to use them here.)

The CCS is the brain of the space-craft. It carries out the instructions sent from the ground to operate the Voyager; it performs the 'housekeeping' functions, and it gathers in the science data. It must also be alert for any problem with, or malfunction of, the numerous sub-systems, so that action can be taken immediately. The CCS consists of two computers, each of which has a 4096 word (18 bit words) memory. Both are used in a non-redundant fashion, but the failure of one of them would not mean total disaster; the Voyager could still operate. The observational plan for the whole encounter period requires about 18 000 CCS words – for example, about 275 CCS words are needed to perform a trajectory correction manœuvre. However, these available words have to be divided between the various time periods of the encounter, which are bound to extend over several months.

There is a two-way communication system with the Voyager space-craft; the uplink contains the command data, while the downlink contains the science and engineering telemetry data. The engineering data are transmitted at 40 bits per second (bps) at the S-band, and are also embedded in the X-band science data, transmitted at higher data-rates – between 4.8 and 21.6 kilobits per second (kbps).

The FDS, which consists of two re-programmable digital computers, is responsible for the collection and formatting of the telemetry data. For the Uranus encounter, it was also responsible for some data processing of the imaging data, through the operation of the data compression algorithms which are needed to reduce the volume of information finally transmitted back to the ground stations. There was a hitch before the Uranus fly-by, when one of the FDS computers lost a 256 memory-block out of the total of 8192 words. This may be regarded as a relatively minor failure, but the loss of a further 512 words from the primary memory would be much more serious, and would mean that the dual-processor mode of the system would have to be abandoned.

There are two classes of space-craft: spin-stabilized and three-axis stabilized. The first of these, used for Pioneers 10 and 11 and for Pioneer Venus, stabilized the space-craft by spinning so that the entire probe acts as a gyroscope. The second method, used with the Voyagers, maintains a fixed orientation in space except when manœuvring. This is achieved by the AACS computer system, which also controls the motion of the scan platform. Gyros may be used for special purposes over short periods of time (a few hours) to maintain the attitude of the probe. In the celestial control mode, Voyager keeps its fixed attitude in space by fixing on the Sun and a bright star such as Canopus, Alkaid, Fomalhaut or Achernar. If the space-craft should drift from its proper orientation by more than an agreed amount, then the AACS will issue commands to fire the hydrazine jets and correct the attitude.

There are times when the Voyager cannot send data directly to Earth – during a manœuvre, for example, or when the probe is behind the target planet. A digital tape recorder (DTR) is then used to store the data and play it back to Earth at a suitable moment. The DTR has eight tracks, each of which can hold an amount of data equivalent to 12 images (6MB). Normally, the DTR is shared between all the science experiments. It is operated at three speeds: 115.6 kbps (record), 21.6 kbps (playback) and 7.2 kbps (record and playback).

The eleven scientific instruments consist of ten experiments and the space-craft radio system. They can be divided into two general classes: particle and field instruments, which are body-fixed, and directional instruments, which are concerned with remote sensing of the atmospheres, satellites and ring systems of the giant planets. There are five directional sensors: imaging science with the narrow- and wide-angle television cameras (ISS), the infra-red interferometer spectrometer radiometer (IRIS), the photopolarimeter (PPS) and the ultra-violet spectrometer (UVS). All these are mounted on the steerable scan platform. The remaining six instruments are used to measure the energetic particles, radio emissions and the magnetic fields of the various planetary systems, as well as in interplanetary space.

There are some important differences between the Voyager imaging system (ISS) and the familiar video camera. The ISS has 800 scan lines, as against the 525 lines of the typical camcorder. While the ISS needs from 48 seconds to several minutes to 'read out' the image, a camcorder can scan an image 60 times each second, which is quite a difference! Moreover, while a video recorder can produce a colour picture very easily, the ISS has to take black and white images through different filters and then use image-processing systems to produce the composite colour image of the target object. Clearly it is all highly complicated, but there can be little doubt that the Voyager instruments represent the most complete set of experiments yet flown in our exploration of the Solar System.

Upgrades for the Neptune encounter

The Neptune 'space spectacular' by Voyager 2 was a major engineering feat, particularly since the probe was relatively old. When it was launched, in 1977, it was designed to operate at peak performance for no more than five years. Yet for the Neptune encounter it had to continue to operate faultlessly for more than twelve years after launch. Indeed it did – and the result was a faultless encounter, making a perfect end to Voyager 2's main career.

However, there had been problems, some of which looked initially ominous. On 5 April 1978 the CCS automatically switched to its back-up receiver. Unfortunately the back-up receiver had concealed a faulty capacitor, so that it was unable to lock-on to the frequency of the transmitted signal. When the prime receiver was switched on, it promptly failed. This means that almost the entire mission has had to be flown by using the faulty receiver; the engineers tune into it with the precision of a short-wave radio – essential, because the exact frequency for communication is very sensitive to the changing temperature of the receiver itself. If the transmitted frequency is not within 96 Hz of the receiver rest frequency, then Voyager 2 will turn a deaf ear to signals from Earth.

Almost a hundred minutes after the closest approach to Saturn, on 26 August 1981, the azimuth motion of Voyager 2's scan platform suddenly jammed. Priceless, unrepeatable observations of the planet, its satellites and the surrounding environment were lost, to the dismay of all those at Mission Control. Apparently the scan platform had seized up during a high-rate slew procedure (about one degree per second); the lubricant migrated away from a tiny shaft-gear interface which was spinning at 170 revolutions per minute, so that the shaft-gear heated up, expanded and stuck. Initially it was feared that the space-craft might have been hit by an icy particle from Saturn's rings. Luckily this was not so, and after two days the scan platform could be slewed again, but it had been an anxious moment, and a detailed investigation of the problem had to be made before the Uranus encounter.

There was plenty of time, because the journey out to Uranus took a further four and a half years. After exhaustive tests, it was found that the scan platform behaved correctly if operated at a slew rate of no more than 0.08 of a degree per second, and in the event there were no problems during the Uranus fly-by. No chances were taken at Neptune. With the exception of the eight medium-rate (0.33 degrees per second) slews to capture some critical science observations, no fast slews were used during the encounter. As a precaution, the engineers had developed a plan for rotating the entire space-craft during the near-encounter phase of the mission – but to the relief of everyone concerned, this did not prove to be necessary.

Since the sunlight at Neptune has an intensity only 1/1000 of that at the Earth, and no more than 40 per cent of that at Uranus, very long exposures were needed to obtain images of the Neptunian system – and it was already known that there were very dark objects and very tenuous ring material. The exposure times ranged between 15 and 96 seconds; real-time images needed exposures in multiples of 48 seconds. To complicate matters still further, the images were always vulnerable to smearing due to the motion of the space-craft, and this meant that a number of important changes had to be made. The Attitude Control system was modified in order to compensate for the impulses generated by stopping and re-starting the tape recorder; this reduced the smearing problem – particularly vital during the closest approaches to Neptune and Triton. Further precautions were also taken, including application of:

The Goldstone radio 'dish', used during the Voyager 2 encounter with Neptune.

(1) the classical image motion compensation method, previously used during the Uranus encounter. This involves rotating the entire space-craft so as to track the target during the exposure. This means moving the Voyager out of contact with Earth and tape-recording the images for later transmission.

(2) the 'nodding' image motion compensation method – similar to the classical technique, but rotating the space-craft only by an amount which makes it possible to keep in touch with Earth. Images obtained in this way are transmitted as they are taken; the space-craft then rolls or 'nods' back to its original position, and there is no loss of contact.

(3) the manœuvreless image motion compensation technique. Only the scan platform, carrying the cameras, is rotated; the attitude of the space-craft itself remains unchanged, so that again there is no loss of contact with Earth.

The clarity of the high-resolution images of Neptune, Triton and the rings shows that these techniques worked well – indeed, better than the planners had dared to hope.

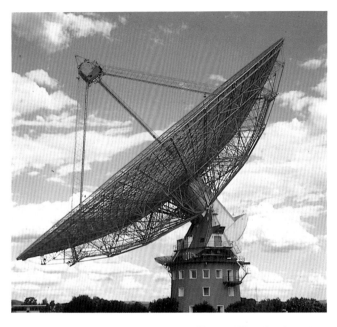

The Parkes radio telescope; and the 'dish' at Tidbinbilla (both in Australia), used during the Voyager 2 pass of Neptune.

The tremendous distance of Neptune from the Earth naturally creates extra problems, simply by decreasing the strength of the radio signals – the further away Voyager is, the weaker its signals must become. At Jupiter the data-rate was 115.2 kbps, but no more than 21.6 kbps at Uranus. In fact, the strength of the signals coming back from Neptune was 36 times weaker than those from Jupiter. Yet the problems caused by the extra distance were overcome. Though Neptune is 1.5 thousand million kilometres further away than Uranus, the data were transmitted at the same data-rate: 21.6 kbps. This was achieved by using on-board space-craft data compression together with significant improvements of the equipment at the Deep Space Network on Earth – a truly remarkable feat.

The special techniques used to send back the data in a more economical manner, without reducing the data-rate, were highly successful. Instead of transmitting the full 8 bits (256 grey levels) of each pixel, or picture element, only the difference between the brightness of successful pixels were transmitted. This reduces the number of bits required for each image by 70 per cent.

Antennæ of the VLA (Very Large Array) at Socorro, New Mexico, used to track Voyager 2 during the Neptune pass.

At NASA's Deep Space Network (DSN), each of the three 64-metre tracking antennæ were enlarged to 70 metres; a high-efficiency 34-metre tracking station was added to the complex at Madrid in Spain, and in Australia the 64-metre antenna and the 64-metre Parkes radio telescope (320 kilometres away) were again linked to form a powerful array. Signals from Voyager were also collected by the twenty-seven 25-metre antennæ of the National Radio Astronomy Observatory's Very Large Array (VLA) in New Mexico; these were equivalent to two 70-metre antennæ, and were combined with the signals received at Goldstone. During the closest approach to Neptune, the 64-metre tracking antenna at the Japanese Institute of Astronautical Science in Usuada was added to the network of ground stations. Certainly this was a truly international programme!

While the engineers had done everything possible to deal with all known problems, there was always the chance of an unexpected crisis at a critical moment. Mission Control, the Jet Propulsion Laboratory in California, is in an earthquake zone, and a sudden shock could well cause a breakdown in communications with the remote DSN sites. If the encounter computer loads had not already been sent up to the space-craft at the time of a major quake, then everything could be lost. To allow for this,

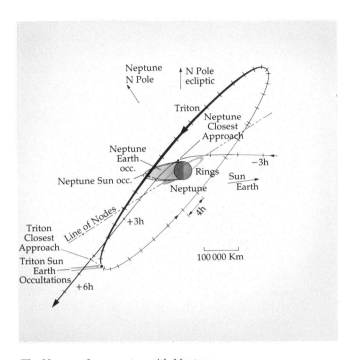

The Voyager 2 encounter with Neptune.

Time on 25 August; the radio signals then took 4 hours 6 minutes to reach the Earth. Closest approach to Triton was at 0910 Universal Time, at a distance of 39 800 kilometres from the satellite. The encounter period formally ended on 2 October, when Voyager had receded to 103 000 000 kilometres from Neptune.

The observational sequences for the encounter were set up through a sequence of ten computer command loads, transmitted to the space-craft during the encounter period. These included both the scientific investigations and the engineering 'housekeeping' measurements which had to be taken regularly for calibration purposes. They covered:

Observatory phase (OB)	5 June–6 August
Far-encounter (FE)	6–24 August
Near-encounter (NE)	24–29 August
Post-encounter (PE)	29 August–2 October
Post-Neptune cruise	2 October–20 November

The design of the Neptune science sequence relied heavily upon telescopic observations from Earth and from the knowledge and experience gained from previous Voyager encounters. The timings of events had to be very precise, since they involved such observations as weather forecasting techniques (to identify previously unknown meteorological features such as cloud systems, in the OB and FE phases, for tracking and morphological investigations), radio and stellar occultations, and searching for new satellites – quite apart from incorporating all the competing observations from the eleven instrument teams.

It all worked well. And we emerged with entirely new knowledge about the fascinating Neptunian system.

copies of all the most critical loads were kept at each of the DSN stations. In the event, there were no sudden crises, which was a great relief.

The Neptune encounter began officially on 5 June 1989, when the space-craft was 117 000 000 kilometres from its target. Closest approach, at 29 240 kilometres from the centre of the planet, was at 0356 Universal

The structure of Neptune

As we have seen, Neptune is the smallest of the gas-giants, though it is considerably more massive than Uranus. The upper cloud-layers are predominantly of hydrogen, with a considerable amount of helium together with hydrogen compounds such as ammonia and methane. It is indeed the methane which explains the characteristic blue colour of the planet, evident even from Earth and vividly shown in the Voyager pictures. Methane absorbs the longer-wavelength regions of the solar spectrum, leaving only the blue.

For our detailed information we depend mainly on Voyager 2, but it would be idle to pretend that our knowledge is complete even now, and there are still discussions about the internal make-up of the globe, and at present there are two main theories.

(1) Beneath the extensive layers of clouds there is an ocean of superheated water, in the region of the mantle lying over an Earth-sized rocky core. This extensive ocean may originate from collisions with thousands of millions of comets during the early life of the Solar System; comets, after all, are made up largely of watery substances. The collisions, together with the high pressure in the area where the ocean now lies, explain the superheated state of the water.

(2) The interior composition is dominated by planetary ices, chiefly water. There may or may not be a rocky core, made up of magnesium silicates and iron; if there is, it may merge in with the icy regions. This would indicate that Neptune (and Uranus too) must have been formed from the accumulation of icy 'planetesimals'.

The second of these models now seems to be the more popular, but we cannot be certain. What is very important, of course, is the amount of internal heat, which is very considerable. In fact, the effective temperature[†] is slightly higher than that of Uranus, even though Neptune is so much further away from the Sun. One major problem now is to explain the lack of an internal heat-source for Uranus, which is certainly the 'odd one out' among the giants – as Table 2 shows.

Since the rotational axis is inclined to the plane of the ecliptic at an angle of 29 degrees (slightly greater than that of the Earth), there will be marked seasons; at the time of the Voyager 2 pass the northern hemisphere was having its winter. We must also remember that Neptune's orbit is appreciably eccentric, and that the difference between the nearest and furthest distances from the Sun amounts to over 80 000 000 kilometres, which must also have an effect.

† The effective temperature is a measure of the energy output of an object, defined as the temperature of a black body having the same total output of the observed output (a black body is a theoretical concept – a body absorbing all the radiation falling upon it). Scientifically, too, temperatures are often given in degrees K, starting at absolute zero: −273 degrees Centigrade. Therefore, the equatorial temperature of Neptune, −226.4 degrees C, corresponds to a temperature of 59.3 K.

Table 2. *Energy balance of the outer planets*

	Jupiter	Saturn	Uranus	Neptune
Rotation period, hours	9.92	10.66	17.24	16.11
Albedo	0.343 ± 0.03	0.342 ± 0.03	0.3 ± 0.05	0.29 ± 0.7
Equatorial temperature, degrees C.	−163.5 ± 1.4	−187.6 ± 0.9	−214.8 ± 1.0	−226.4 ± 1.1
Effective temperature, degrees C.	−148.6 ± 0.3	−178.0 ± 0.4	−213.9 ± 0.3	−213.7 ± 0.8
Energy Balance	1.73 ± 0.16	1.84 ± 0.17	1.03 ± 0.09	2.61 ± 2.8

The atmosphere of Neptune

Let us now turn to the only part of Neptune which we can actually see directly – its atmosphere. Overall, the atmospheres of the giant planets are hydrogen-rich reducing envelopes, quite unlike the oxidizing atmospheres of the terrestrial planets. In the case of Neptune, the upper atmosphere has been found to consist of 85 per cent hydrogen, 13 per cent helium and 1–2 per cent methane (as we have seen, it is methane which accounts for the planet's blue colour). The ultra-violet spectrum of the sunlit hemisphere is very similar to the spectra of Jupiter, Saturn and Uranus.

We have not yet been able to identify many constituents in the atmosphere of Neptune, either from ground-based observations or from the Voyager encounter. Remember that although Voyager 2 flew close to the planet, most of the instruments on the scan platform – and therefore responsible for the remote sensing observations – had not been specifically designed to carry out studies of Neptune in the cold, distant reaches of the Solar System, where the signals are so much weaker than those obtained at Jupiter at the start of Voyager's main mission. The IRIS observations were made with a spectral resolution of so low a value that as many as one hundred spectra had to be combined in order to provide enough power for proper analysis to be carried out.

There is evidence for the presence of methane, ethane, acetylene, carbon monoxide and hydrogen cyanide in Neptune's atmosphere. Further derivatives of methane are also to be expected. The concentrations will vary through the troposphere and stratosphere as these constituents are involved in the photochemical processes which modify the atmospheric composition and create local aerosols. Apparently there is a regular cycle of events. The ultra-violet radiation from the Sun destroys the methane in the planet's upper atmosphere, converting it to hydrocarbons which descend into the cold lower stratosphere; here they evaporate, and then condense. The hydrocarbon ice particles fall into the warmer troposphere, evaporate, and are then converted back to methane and methane clouds which rise into the stratosphere and replenish the region with ethane. This cycle means that the amount of methane is kept more or less constant.

The other minor constituents are more variable – acetylene, for example. It seems that at the time of the Voyager encounter, most of the acetylene emission originated in the region extending from 0.03–2.5 millibars, with maximum concentration at the 1.5 millibars level.

In the troposphere we can expect variable amounts of hydrogen sulphide, methane and ammonia, all of which are involved in the creation of cloud layers and associated photochemical processes. There are two possible ways to account for the presence of carbon monoxide and hydrogen cyanide. One is by rapid transport from the deep troposphere, though this may be unlikely in view of the fact that the observed amount of hydrogen cyanide seems to be too great to be explained in such a way. Alternatively, carbon monoxide and hydrogen cyanide could be created in the rich and dynamic chemistry of Neptune's upper atmosphere, and transported into the 'weather' regions. Certainly the hydrogen cyanide could be produced in the stratosphere by photochemical reactions involving the methane hydrocarbon products together with nitrogen. It may even be that the nitrogen could be sent into Neptune's atmosphere from Triton, which does have a nitrogen atmosphere (albeit a tenuous one). The carbon monoxide could be formed photochemically from the conversion of water provided by comets. The precise composition of the deep atmosphere is unknown, but it has been suggested that there may be a much greater proportion of carbon in Neptune's atmosphere as compared with the solar mixture – perhaps by a factor as great as ten.

Atmospheric structure

There are marked similarities between the compositions of the reducing atmospheres of the giant planets. However, now that Neptune has been studied from Voyager 2, there is no doubt that Uranus is the strange member of the family, with its lack of internal heat and its bland appearance – possibly because any major atmospheric features are submerged beneath the ubiquitous aerosol layers.

Although the solar heating is greatest in the equatorial region of Neptune and the polar regions of Uranus, the horizontal temperature structures of the two planets are much the same. In each case the poles and equatorial regions have approximately the same local temperature, while the mid-latitude regions are a few degrees cooler. In this part of Neptune, the minimum temperature at the 100 millibar pressure level is about −223 degrees C (50K).

Variations in heating and photochemical processes mean that the structure of the atmosphere in the stratosphere, and above, is locally changeable. The temperatures in the upper atmosphere are determined by the ionization of the molecular hydrogen, though in the troposphere the structure tends to follow the adiabatic lapse-rate in the usual way.

There are, however, significant zonal variations in the temperature structure of Neptune, with warm bands, cold bands, and discrete temperature features which are in general correlated with the cloud features which we can observe. This situation is in

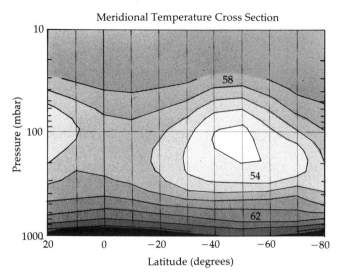

Meridional cross-sections of temperature in the atmosphere of Neptune. From the Voyager observations of B. Conrath, F. M. Flaser and P. Gierasch (1991) J. Geophysical Res. 96, 18931–40.

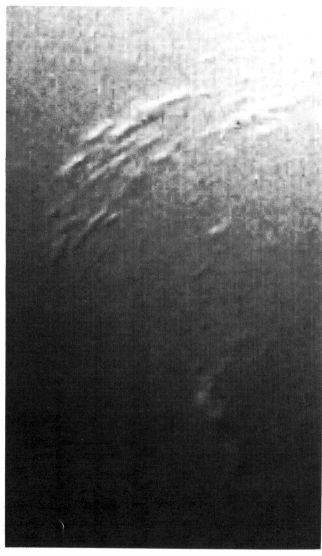

Neptune's South Polar Region: Voyager 2's narrow-angle camera; 23 August 1989, from 2 570 000 km. The smallest cloud features are 45 km in diameter. The image shows, for the first time, shadows in Neptune's atmosphere – shadows cast on to a deep cloud-bank by small, higher clouds. Latitude 68 degrees S.

sharp contrast to that at Uranus – or, indeed, at Saturn, and all in all there is a greater similarity between Neptune and Jupiter. Zonal temperature observations indicate that on Neptune there is a relatively warm region round the pole and another round the equator, while the mid-latitudes are cooler. In the stratosphere, the temperatures range between −215 degrees C (58K) at the equator to −220 degrees C (53K) in mid-latitudes. Smog layers are found in the lower stratosphere and upper troposphere, extending with variable composition and opacity from the 5 millibar level. From what we know about the photochemistry of methane, we may expect to find lower-order hydrocarbons such as ethane, acetylene and diacetylene. In the troposphere there are several distinct cloud layers – methane hazes and clouds at about the 1.5 bar level, with more opaque cloud at the

3.3 bar level made up of hydrogen sulphide ice particles, possibly with some ammonia also.

Clouds, hazes and atmospheric motions

Before the Voyager fly-by we had very little information about Neptunian meteorology. From Earth, Neptune's disk is only slightly larger than that of Jupiter's satellite Ganymede. The best pre-Voyager results were probably those of B. A. Smith and R.

Facing page
South Polar Region of Neptune, from 900 000 km; 27 August 1989. The resolution is 120 km. Near the bright limb, clouds at latitudes 71 and 42 degrees S, are shown; a bright cloud at the bottom left lies within 1.5 degrees of Neptune's South Pole.

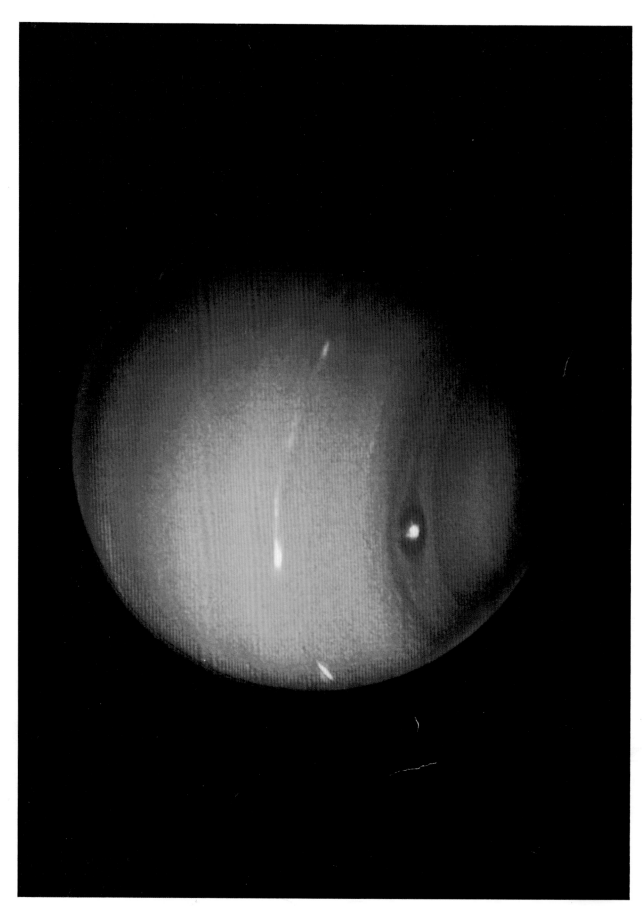

False-colour image of Neptune; 26 August 1989. The picture was made from images taken through the blue and green filters, and a filter which passes light at a wavelength which is absorbed by methane gas.

Terrile, using a CCD (Charge-Coupled Device) on the 2.5-metre Irénée du Pont telescope at the Las Campanas Observatory in Chile, and those of Hammel from Hawaii. However, all that could be found was evidence of some discrete, variable cloud features. Neptune certainly seemed to have an active, dynamic atmosphere, but until Voyager neared its target we had no positive information. We now know that Neptune does indeed show a wide range of cloud features – as the Earth-based data had so tantalisingly suggested.

Neptune's bluish atmosphere is dominated by a large anti-cyclonic storm system, now called the Great Dark Spot (GDS), situated at latitude 20 degrees South. The GDS is about the same size as the Earth, with an average extent of 38 degrees in longitude and 16 degrees in latitude; in some ways it is not unlike the Great Red Spot on Jupiter, where the extensions are to 30 and 20 degrees respectively. The GDS is actually 'less blue' than the surrounding region; relative to Neptune, its size is about the same as the Great Red Spot relative to Jupiter. It rotates anti-cyclonically in a period of 18.2 hours, and drifts westward at about 30 metres per second relative to the adjacent clouds. In many ways it behaves rather like a ball-bearing! It is about 10 per cent darker than its surroundings, while the nearby material is 30 per cent brighter – indicative of the difference in altitude between the two features. Certainly the GDS has all the characteristics of an atmospheric vortex.

Spots on Neptune; 23 August 1989, from 4 200 000 km. The smallest features visible are 38 km across.

The Great Dark Spot, from 2 800 000 km; resolution 50 km. The image was secured with the narrow-angle camera.

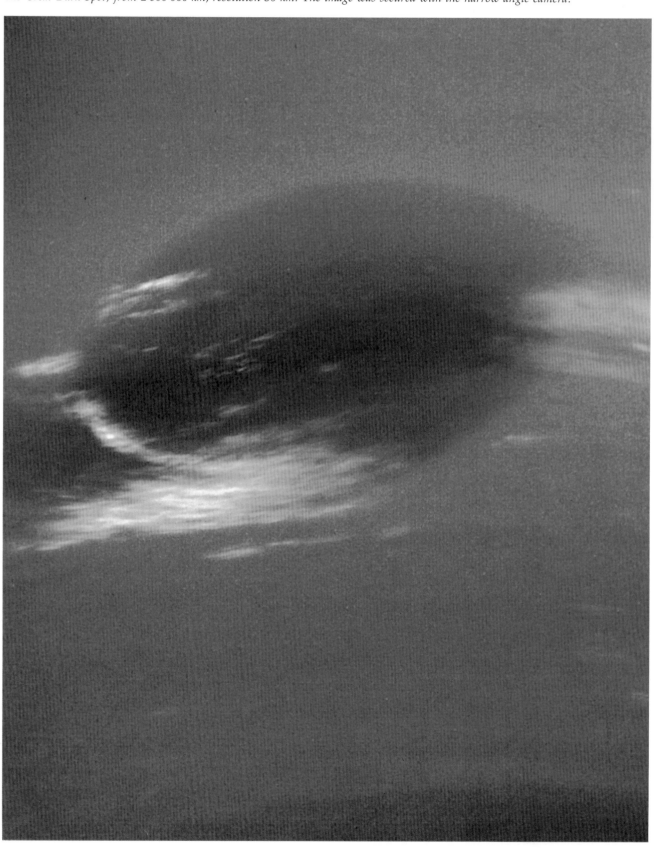

There are, however, some significant differences between the GDS on Neptune and the Great Red Spot on Jupiter. The circulation of the Red Spot modifies the surrounding environment, forming a turbulent wake to the west. While Neptune's GDS also affects the immediate environment, the surrounding regions do seem to be much more uniform than those on Jupiter. Moreover, the westward drift of the Red Spot is a mere 3 metres per second, virtually matching the planet's rotation. The rapid motion of the GDS on Neptune relative to its surroundings may simply indicate that it is deep, while Jupiter's Red Spot is shallow. Furthermore, for the easterlies on Jupiter and Saturn the driving layer for the motion may be closer to the surface than on Neptune.

The bright, cirrus-like companion along the southern edge of the GDS was detected in the Voyager images in January 1989. While this was the first discrete feature seen from Voyager 2, it cannot be claimed that it has been positively identified in Earth-based pictures, though the best images obtained from the ground do show a bright feature at latitude 33 degrees South which is probably associated with the southern edge of the GDS. Similar features have been seen at 33 degrees South in 1988 and at 38 degrees South in 1986 and 1987. The variations in the shape and brightness of this feature are similar to those of orographic clouds, such as the lenticular clouds found in the atmospheres of the Earth and Mars – which are produced when air is forced upward and over a ridge or mountain system. Though we do not have such rigid features in the atmosphere of Neptune, the local pressure and temperature anomalies around the GDS may play a similar rôle.

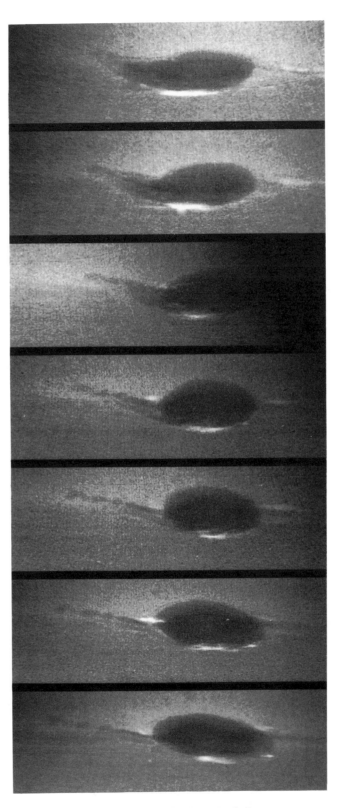

Changes in the clouds around the Great Dark Spot, over a period of 4.5 days. The violet filter of the narrow-angle camera was used to produce these images at distances ranging from 17 000 000 km at the top to 10 000 000 km at the bottom.

Left
Clouds on Neptune; 21 August 1989, from 16 000 000 km. The series of six images shows altitude differences in the clouds. The top three images, taken through orange, violet and ultra-violet filters in the narrow-angle camera, show bright cloud features. The 'Scooter', prominent in orange, is invisible in ultra-violet, where scattering by atmospheric molecules is strongest. The lower three wide-angle camera images [left to right: orange, weak methane (541 mm) and strong methane (619 mm)] are arranged in order of increasing absorption by methane in Neptune's atmosphere. The Scooter is less obvious from left to right, implying that there is relatively more absorbing methane above it, so that it lies deeper down than the other bright colours.

Methane cirrus clouds on Neptune; 25 August 1989. This image was taken only 2 hours before Voyager 2's closest approach to Neptune. The range was 157 000 km; resolution, 11 km. The bright sides of these long clouds face the Sun, while shadows can be seen on the sides away from the Sun. The clouds are around 50–200 km wide, and the widths of the shadows range from 30–50 km. The heights of the clouds are about 50 km.

In April 1989 a bright feature was detected round the southern pole, at latitude 71 degrees South. This south polar feature (SPF) is not a single cloud system, but an active arc extending over more than 90 degrees of longitude, though restricted to a latitude band of less than 5 degrees; the height above the cloud deck is about 50 kilometres. A small feature at the south pole of rotation suggests that there is a well-organized polar circulation that is quite unlike the situation in the polar regions of the other giant planets. Jupiter and Saturn have pronounced polar hazes; Uranus and Neptune do not, but they do show dark hoods at ultra-violet wavelengths, which may be the result of the production of photochemical aerosols through auroral bombardment. This process would not operate on Uranus or Neptune if the methane concentration in the atmosphere were too deep to be affected by the particle heating process.

Other similar cirrus-like cloud features are seen at latitude 27 degrees North, while bands of lower reflectivity extend from 6–25 degrees North and from 45–70 degrees South latitude; all these seem to lie above the methane cloud layer. The streaks at 27 degrees North were seen to cast shadows on the cloud deck, which was thought to be from 50–100 kilometres below. The widths of the streaks ranged from 50–200 kilometres, while the width of the shadows ranged between 30 and 50 kilometres.

The third bright feature discovered by Voyager has

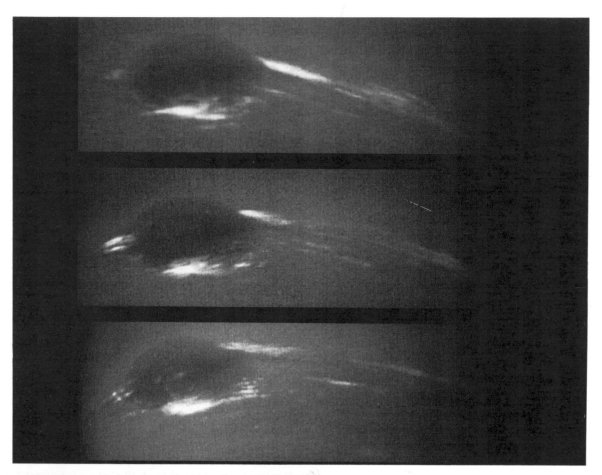

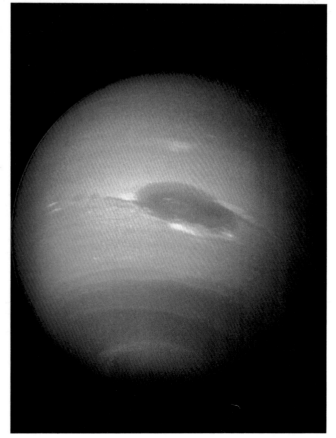

Cirrus clouds over the Great Dark Spot. The sequence spans two rotations of Neptune; rapid changes are evident.

False-colour image of the Great Dark Spot; 21 August 1989, produced from images taken with the ultra-violet, violet and green filters of the wide-angle camera. The GDS has a deep-bluish cast, indicating that visible light (but not ultra-violet) can here penetrate to a deeper layer of dark cloud or haze in Neptune's atmosphere.

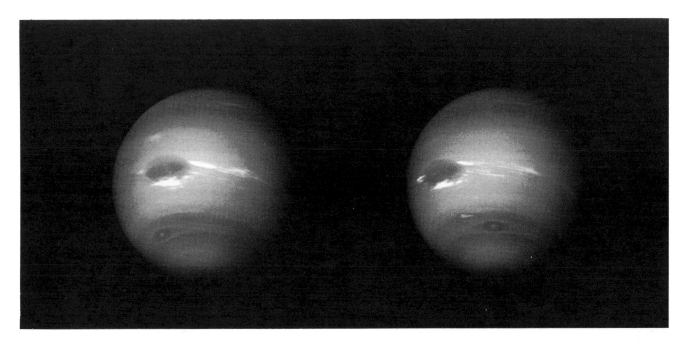

Neptune, showing the Great Dark Spot, 21 August 1989, from 12 000 000 km. The resolution is 110 km per pixel. During the 17.6 hours between the left and right images, the GDS had completed a little less than one rotation of Neptune; spot D2 had completed a little more than one rotation, as can be seen by comparing its relative position in the two pictures.

been nicknamed 'the Scooter'. It is bright and compact, and lies at about latitude 42 degrees South. When it was first observed, the precise rotation period of Neptune was unknown, and by adjusting the time-lapse sequence to a period of 17.5 hours – which was then the best estimate, but now known to be too long – the Scooter appeared to be moving much more quickly than it really does. In spite of this, the name has been kept. The Scooter is deeper in the atmosphere than the cirrus veils, and may be an upward extension of the deeper cloud-deck.

A small dark spot (D2), with a bright core, is located at latitude 55 degrees South. During the Voyager encounter period the spot's core was seen to increase in brightness, and some small-scale features were clearly visible inside the spot itself. The sizes and details of these features changed markedly over periods of hours. Features of similar type have been seen in the Jovian atmosphere, and have been found to rotate anti-cyclonically, but we cannot yet be sure of the direction of rotation in Neptune's D2.

Even though Neptune receives only 3 per cent of the sunlight available at Jupiter, and a mere 0.1 per cent of that on Earth, Neptune is a surprisingly dynamic planet. At Jupiter and Saturn, the wind speeds and directions were derived from tracking small features over short time-intervals, careful allowance being made to ensure that there was no morphological change to the cloud systems over this period. On Neptune, however, the small-scale features evolve rapidly and disappear quickly, so that considerable care must be taken in the process of measurement and interpretation.

The large-scale atmospheric features move at wind speeds ranging from 20 metres per second (prograde) at latitude 54 degrees South to 325 metres per second (retrograde) at latitude 22 degrees South.[†] This makes Neptune the windiest planet in the Solar System; the windspeeds approach the velocity of sound, which, at a temperature of −213 degrees C (60K), is 560 metres per second. The rotation period is about 19.5 hours at the equator; this is longer than the radio period of 16.11 hours and the corresponding period of about 16 hours for the high latitude regions of the planet. A broad equatorial retrograde jet extends from approximately latitude 50 degrees South to at least 45 degrees North. A relatively narrow prograde jet of at least 300 metres per second is found at latitude 70 degrees South.

But why should Neptune, and indeed Uranus, have super-rotating atmospheres? Uranus, so similar to Neptune in size, has a similar atmospheric circulation, with a super-rotating atmosphere poleward of 20 degrees South and a sub-rotating atmosphere equatorward of that latitude. On Neptune, the sub-rotating atmosphere is in a wider belt between latitudes 54 degrees North and South. Perhaps Neptune's rapid retrograde winds are maintained by upwelling from the deep interior, as Suomi and his colleagues at the University of Madison suggest; this could account for extra differences in the motions that we observe. This plausible mechanism is based upon

† *Prograde* indicates an eastward wind, in the same direction as the planet's rotation; *retrograde* winds are westward, opposite to the direction of rotation.

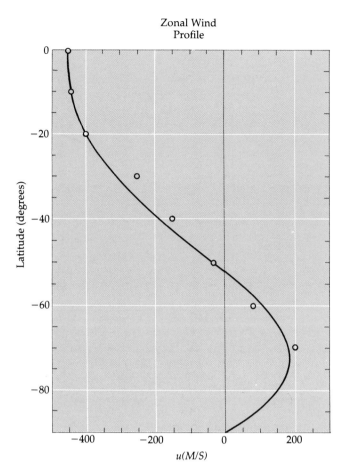

Zonal Wind Profile

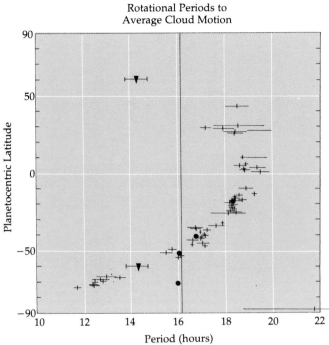

Rotational Periods to Average Cloud Motion

Zonal wind profile on Neptune. [From the article by G. Conrath, F. M. Flaser and P. Gierasch (1991), J. Geophysical Res. **96** *18931–40.]*

the principles of angular momentum and energy conservation in conjunction with deep convection, which then leads to a régime of uniform angular momentum at low latitudes. The rapid retrograde winds would then be the result of deep convection and the efficiency of Neptune's heat engine. It follows that Neptune transfers more energy from its interior, and supports deeper convection, than with Uranus. However, the most remarkable aspect of Neptune's zonal wind profile is its similarity to that of Uranus. The two planets are completely different with regard to their energy sources and axial tilt, so that it is very surprising to find that the zonal wind systems are so alike.

Extended atmosphere and magnetospheric interactions

Rather weak auroral emissions (5×10^7 watts) have been observed on the dark side of the planet near longitudes 30 and 200 degrees. It is suggested that the aurora is driven by a plasma arc escaping from Triton's atmosphere and then becoming ionized. Since the magnetosphere of Neptune is sparsely populated, weak emissions are to be expected.

In the region of the thermosphere, the observed temperature is about 477 degrees C (750 K), ±150 degrees. This value is close to that found on Uranus,

but Neptune's higher gravity and colder stratosphere mean that there are much lower number densities in Neptune's upper atmosphere than in that of Uranus.

Ionosphere

Above the mesosphere is the ionosphere, where the density is very low and the electrical conductivity increases with altitude. The name is derived from the higher proportion of atoms and molecules which are ionized.

The Voyager 2 observations at 3.6 and 13 centimetre wavelengths provided the first evidence of the Neptunian ionosphere in a region between 61 degrees North and 44 degrees South. The ionosphere is extensive, with a topside temperature of about 690 degrees C (960 K), ±160 degrees, assuming that hydrogen is the dominant ion. There are well-separated, narrow ionospheric layers, which are similar to those of Jupiter, Saturn and Uranus, and may be similar to the sporadic E layers on Earth. The layers lie at altitudes of 650, 690, 725 and 750 kilometres relative to the 1-bar level. By analogy with the sporadic E layers in the Earth's ionosphere, it seems that the Neptunian layers are probably composed of metallic ions produced during meteorite ablation. It is also possible that ring particles could be a further source of material.

The magnetosphere of Neptune

The possibility of a magnetic field and magnetosphere associated with Neptune had been under discussion for many years before the Voyager mission. In 1988 came the detection of possible synchrotron radiation near the planet, which suggested that there might be a magnetic field with a surface field strength between 0.3 and 17 gauss (the Earth's field at the surface has a strength of 0.5 gauss). However, at that stage we had no information about the nature and extent of the magnetosphere.

A magnetosphere is a windsock-shaped region surrounding a planet, with the head directed toward the Sun and the tail streaming out behind in the solar wind. It is due to energetic particles trapped within a planet's field lines. The different planets have very different magnetic properties. Mercury has a massive iron core but a weak field; Venus and Mars have virtually no fields at all, and neither has the Moon. The Earth, Jupiter and Saturn have substantial magnetospheres, and in fact Jupiter's is so large that if it were visible it would have the same angular diameter as the Sun. Since there are many similarities between the outer planets with regard to composition and quick axial rotation, it seemed reasonable to invoke a 'magnetic Bode's Law' which would indicate the presence of a strong field and extensive magnetosphere for Neptune. Furthermore, the initial explanation of the 59-degree tilt of Uranus' magnetic field with respect to the axis of rotation was associated with the planet's unusual tilt and the fact that Voyager 2 flew past at the time of a possible dynamo reversal. There was no reason to expect that Neptune would be behaving in the same way. It was felt that Neptune would follow the standard pattern, so that Voyager would pass through the auroral region at the north pole while sampling the properties of the magnetosphere. But as Neptune started to unveil its secrets, it became clear that we were due for a major surprise.

Magnetic Field

Neptune has a magnetic field which is tilted by 47 degrees to the planet's rotational axis, and is offset by $0.55\ R_N$ which is about 13 500 kilometres.[†] It was only when the tilt of the field was discovered that the encounter observations through the magnetosphere and over the pole started to make sense.

The first direct evidence of the magnetic field of Neptune came from the detection of radio emissions, eight days before closest approach, when the spacecraft was still $470\ R_N$ (11 750 000 kilometres) from the planet. Subsequently, Voyager crossed the bow shock at $34.9\ R_N$ (872 500 kilometres); the bow shock marks the boundary between the solar wind and the planet's magnetosphere, creating a shock front which resembles that made by an ocean liner on the surface

† R_N = the radius of Neptune: approximately 25 000 kilometres.

The magnetosphere of Neptune at the time of the Voyager 2 encounter. The prominent features are the radiation belts confined to the orbit of Triton, the compressed field on the day side and the magnetic tail streaming out at a great distance from the planet.

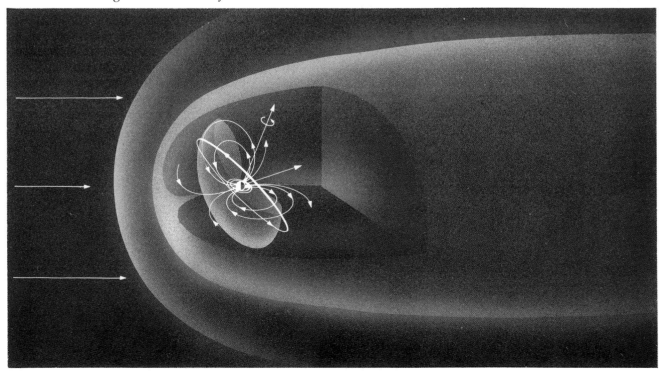

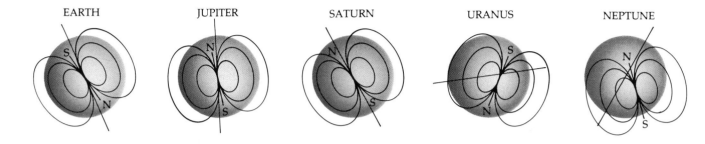

Relationships between magnetic and rotational axes of the planets.

of the sea. Unexpectedly, the inbound magnetosphere was less well-defined. We now know that this was because Voyager had entered a highly tilted magnetic field at a very high magnetic latitude. At least this provided the bonus of the first direct observations of a pole-on magnetosphere, where the solar wind is incident upon the magnetic polar region rather than the equator. Between 4 and 15 R_N 100 000 and 375 000 kilometres) Voyager passed through a region which may be characterized by a strange magnetic dipole field with an axis tilted at 47 degrees and offset by 0.55 R_N from the centre of the planet. The dipole moment is 0.14 gauss, which corresponds to a surface magnetic field which is very asymmetric and which varies between 1.2 gauss in the southern hemisphere to only 0.06 gauss in the northern. The field polarity is the same as those of Jupiter and Saturn, but opposite to that of the Earth. The offset nature of Neptune's field suggests that the dynamo electric currents are probably reasonably close to the surface of the planet rather than near the core. At the time of the encounter, Voyager found that the outbound magnetosphere was located at about 72.3 R_N (1 807 000 kilometres), and the space-craft experienced multiple bow-shock crossings near 161 R_N (4 025 000 kilometres).

A similar analysis of the magnetic field of Uranus yields a dipole moment of 0.23 gauss, displaced along the rotational axis by 0.33 R_U and inclined to the rotation axis by 60 degrees. All the other planets have magnetic axes which are not very different from the axes of rotation, so that in this respect Uranus and Neptune are very alike – and equally unlike all the other planets which have magnetic fields.

The rotation period of the magnetic field, measured from the periodicity of the radio emissions, is 16.11 ± 0.05 hours. Now, at last, we have been able to measure the length of Neptune's day. This corresponds to the rotation period of the deep interior of the planet, and therefore the solid body of Neptune. Therefore, all the meteorological wind observations are measured relative to this rotation period.

As the magnetic field rotates with the planet in 16.11 hours, the satellites and ring particles are swept through this highly charged region, which is bound to affect their basic properties and surface chemistry. As Neptune rotates, the magnetosphere configuration changes from pole-on, with a cylindrical magnetotail, to a more normal configuration. This rapidly changing geometry caused regular changes in the characteristics of the magnetosphere observed by Voyager during the fly-by.

The magnetic field is created by a dynamo action in the deep interior of the planet. The location of this dynamo region may be determined from the need for the presence of an electrically conducting fluid at this level. The conductivity in the upper hydrogen layer must be too small to create the dynamo, but the required conditions can be found at an intermediate depth in the oceanic layer, where the high temperatures and pressures in the huge water ocean are able to generate the observed field because they are sufficiently conducting. (This assumes that the ocean theory for Neptune's interior is correct; if not, we may have to think again.)

As expected, Neptune is a source of radio waves. Lightning has been detected from charged particle discharges, similar to those found for the other giant planets and also in the upper atmosphere of the Earth. Unfortunately, the lightning has not been seen in Neptune's weather systems, and it seems that the lightning storms have only about the same power as those of the Earth. The weakness of the Neptunian lightning may be due to the reduced concentration of ammonia in the planet's atmosphere. It is worth noting that only at Jupiter has there been any visible evidence of lightning in the clouds of any world other than our own.

The fact that Uranus and Neptune have tilted, off-set magnetic fields suggests that these are characteristic of the present state and composition of their interiors. Table 3 summarizes some of the magnetic properties of the planets.

Though the fields of Uranus and Neptune are similar with respect to dipole tilt and offset, the strengths of the dipole moments are surprisingly different; that of Neptune has a value of 2×10^{27} gauss per cubic centimetre, while for Uranus the value is 3.8×10^{27}. Since the two bodies are of similar size and composition, it is reasonable to assume that the dipole moment depends upon the energy available to power the dynamo. Neptune has a much stronger internal heat-source than Uranus (if, indeed, Uranus has any at all), so that the difference between the dipole moments provides further evidence that Neptune's field is much more complicated than that of Uranus. Indeed, it appears that the Neptunian field is best represented by a quadrupole or even an octupole structure. The revelation that the two fields are so similar in some ways and so different in others means that we must now re-examine the conclusions drawn after the 1986 Voyager 2 encounter with Uranus.

It was first thought that the large inclination of Uranus' magnetic field was simply due to the planet's unique axial tilt, but this does not, of course, apply to Neptune. Another idea was that with Uranus we were studying it at the time of a reversal of polarity. In this case the same would be true of Neptune, and both fields would be in what could be called a flip-flop motion. This sort of situation is not unusual – the Earth's field has reversed at least nine times during the past 3 500 000 years, and metal particles in the rocks on the sea-bed have been found pointing in different directions which correspond to different magnetic epochs – but we would then have to assume that Uranus and Neptune were going through polarity reversals at exactly the same time, which seems rather too much of a coincidence. At present we have to admit that we do not know the answer.

Neptune, like Uranus, has an average density much greater than would be expected for a condensed object of composition similar to that of the Sun, or the gas-giants Jupiter and Saturn. It may well be that the interiors of Uranus and Neptune, where the electrical dynamos are generated, are made up of a fluid and convecting 'ice mantle' of water, ammonia and methane, enriched with appreciable 'rock', overlying a small rocky core about the same mass as the Earth. Certainly there must be some major differences between the internal state and composition of Neptune and Uranus, but it seems that in an astrophysical context the skewed magnetic fields may be common to a class of stars known as oblique rotators.

Magnetospheric interactions

Neptune's rings, and all the satellites apart from Nereid, move in the hostile charged-particle environment of the magnetosphere. This is not a new situation. We already knew that at Jupiter, Saturn and Uranus there are important interactions between the charged particles and the ring and satellite systems. Triton has an extended atmosphere and ionosphere, and injects neutral atoms as well as charged particles into the magnetosphere, producing a neutral torus along Triton's orbit. If the magnetic dipole axis of Neptune were aligned with the planet's spin axis, ionization of these neutral particles would form a plasma torus similar to the Io torus round Jupiter (bearing in mind that Io, unlike Triton, has violently active volcanoes which are erupting all the time). But, of course, this is not so; there is a tilt of 47 degrees between the magnetic and the spin axes, so that the Triton torus has a thickness of $2 R_N$ (50 000 kilometres) and an inner boundary at $8 R_N$ (200 000 kilometres). The neutral hydrogen torus density is estimated to be 300 per cubic centimetre. The large tilt of the magnetic dipole also causes the configuration of the magnetosphere to change diurnally, as with Uranus, but – as we have seen – the unusual

Table 3. *Some magnetic properties of the planets*

Planet	Tilt, degrees	Dipole equatorial magnetic field, gauss	Size of magnetosphere, planetary radii
Mercury	14	0.0033	1.4
Earth	11.7	0.31	10.4
Jupiter	−9.6	4.28	65
Saturn	0	0.21	20
Uranus	−59	0.23	18
Neptune	−47	0.1	26

geometry of Neptune's field means that the configuration of the magnetosphere changes from Earth-like to pole-on, and back again, every Neptunian day.

The plasma in the magnetosphere is made up chiefly of hydrogen ions, electrons and some nitrogen ions. Despite the presence of a powerful plasma source, it is the most sparsely populated magnetosphere found during the Voyager missions, with a maximum plasma density of 1.4 per cubic centimetre. Most of the plasma is concentrated in a sheet near the planet. The light ions probably come from Neptune's atmosphere, while the heavier ions – in particular, those of nitrogen – probably escape from Triton.

The trapped particle population includes atomic hydrogen, singly ionized molecular hydrogen, and helium. The estimated relative abundances are 1300: 1: 0.1. The fact that there is a steep gradient of high-energy electrons and ions around 14.4 R_N (360 000 kilometres) suggests that Triton may be responsible for controlling the outer region of Neptune's magnetosphere. The loss of plasma material from the magnetosphere may be due to interactions with the satellites, rings and other materials in Neptune's coronal region, together with precipitation into Neptune's atmosphere.

Neptune, like Uranus, has some of the darkest-known satellites and ring systems; with some bodies the albedo is as low as 2–5 per cent. This may be due to the constant bombardment by charged particles to an extent varying according to the surface properties of the target objects.

Auroral activity, similar to the northern and southern lights of Earth, was observed in Neptune's atmosphere. Terrestrial auroræ are caused by energetic particles spiralling down the field lines in the polar regions. On Neptune, however, the complex magnetic field means that the auroræ are spread over wide regions of the planet, and are not confined to the polar zones. The auroral power of Neptune is weak, amounting to only about 5×10^7 watts compared with 10^{11} watts for Earth. Auroral activity was

also found at Triton, probably due to charged particles from Neptune's radiation belts plunging into Triton's atmosphere. At the orbit of Triton the electron fluxes represent a power input into its atmosphere of 10^9 watts, which is consistent with the observed ultra-violet auroral emission.

Radio emissions

Although the radio bursts from Neptune were first detected eight days before closest encounter, subsequent analysis of the data showed that they could have been detected as early as 30 days before Voyager 2 passed by the planet. These highly polarized emissions occur in the frequency range between 100 and 1300 kilohertz, with an unchanging period of 16.11 hours, and were thought to originate at the southern magnetic pole. Other emissions have also been detected, in the 20–865 kilohertz range, with the same period of 16.11 hours, although a phase shift of 180 degrees was noted as the space-craft moved across the planet.

In contrast to the Saturn and Uranus encounters, there was no evidence of electrostatic discharges associated with the particle interactions in the magnetosphere – perhaps because of the low particle population. Lightning-generated whistlers were detected in the magnetosphere near the magnetic equator, at distances ranging from 1.3–1.99 R_N (32 500–49 700 kilometres). These discharges suggest that Neptune's lightning power is similar to that of the Earth, and could be the result of the lower concentration of ammonia in the planet's atmosphere. The limited evidence for lightning may be due to one of several causes – such as reduced particle-charging rates, very brief duration events, or slow rise-times of the discharges. However, it seems that the type of lightning seen at Saturn and Uranus, involving interaction with the ring particles, does not appear to be present at Neptune.

Rings of Neptune

Neptune has a system of rings. To many people this may not seem to be startling, but it is actually one of the most important discoveries made by the amazing Voyager encounter with Neptune.

Earlier Voyager observations had led to the detection of the tenuous, diffuse ring round Jupiter, had provided a detailed picture of the extensive ring system of Saturn – providing some clues about their stability from the evidence of tiny shepherd satellites – and then obtained close-range observations of the dark, multi-ring system surrounding Uranus. Yet any unambiguous information about the possible Neptunian system was still lacking. Indeed, it is only during the past ten years that there has been the slightest indication that Neptune could have rings of any sort. It had been suggested that they would be absent, or at best only partial arcs because of the gravitational perturbations due to Neptune's satellites, the retrograde Triton and the eccentrically moving Nereid. The only evidence came from occultations of stars by Neptune. There had been ten such events, and on eight occasions there were signs of secondary occultations of the same type as those which had led to the discovery of the rings of Uranus in 1977. The light of the occulted star faded briefly either before or after the occultation by the planet itself.

The dimming of a star's light from such an occultation usually indicates the presence of ring structure, since a solid satellite would cut out the starlight completely. However, with a continuous ring system the dimming would occur both before and after the actual occultation of the star by the planet. It was felt that these unexpected observations might be explained by the presence of narrow ring arcs orbiting between 41 000 and 67 000 kilometres from Neptune, extending 4, 4 and 10 degrees in ring longitude rather than making up a continuous ring. We now know that the arc theory explains all but one of the confirmed ring occultation events seen from Earth. The remaining one was actually a chance observation of the satellite Larissa, the second of the new satellites discovered during the Voyager fly-by.

This confused situation was the ideal setting for the final Voyager encounter, which could produce a definitive answer to this perplexing and fundamental problem but would certainly give rise to a whole new set of problems!

Ring Properties

The data are summarised in Table 4. The Neptunian system observed during the Voyager mission contains:

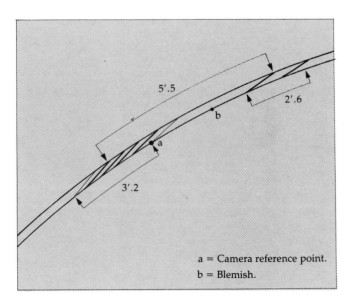

Tantalising images that give a hint of material in the rings of Neptune. The hatched area covers a region of six clumps; the lower hatched region has two clumps in the ring systems.

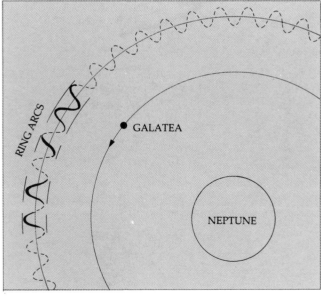

The interactions between the small moon Galatea and the ring particles not only maintain the 'arc' positions but produce the scalloped wave in the Neptune rings. If you were to track the arcs as they orbited the planet, it would appear as if a wave through the arcs at Galatea's orbital speed.

Two narrow rings, Adams and Le Verrier, at radial distances of 62 900 and 53 200 kilometres respectively. The Adams ring has a reddish colour, not unlike that of Jupiter's ring and the Jovian satellite.

A broad ring, Galle, at a radial distance of 41 900 kilometres.

A second broad ring, the 'Plateau', extending outward from the Le Verrier ring to a distance of nearly 59 000 kilometres.

A narrow ring, just closer-in than the Adams ring, which has a radial distance much the same as that of the new satellite Galatea.

A sheet of material which may fill the inner Neptunian system.

The dust content of the Le Verrier and Galle rings is about twice that of the other rings, and more than with the main rings of Saturn and Uranus. The Adams ring is the outermost member of the system, and includes three arcs of denser material. There appears to be more dust in the three ring arcs than in the rest of the Adams ring. The bright ring arcs have been named Liberté, Egalité and Fraternité.[†] These arcs are clustered together within a range of 33 degrees in longitude. The Adams and Le Verrier rings lie about 1000 kilometres outside the newly discovered satellites Galatea and Despina respectively.

† These names were allotted at the 1991 General Assembly of the International Astronomical Union, held at Buenes Aires in Argentina. With their political overtones, they appear to be most unsuitable. One of us (P.M) was present at the Assembly, and proposed that the arcs should be named in honour of D'Arrest, Bouvard and Bessel, all of whom were associated with the discovery of Neptune, but the proposal was not accepted. PM has now made the official proposal that the latest-discovered ring, moving in the orbit of Galatea, should be named the D'Arrest ring. At the time of writing this proposal is still under consideration.

Rings of Neptune (Voyager 2); they are here shown backlit by the Sun. The image of the planet is greatly over-exposed, to capture detail in the rings. The rings appear bright, as microscopic ring particles scatter sunlight toward the camera.

Table 4. *Neptune ring data*

Feature		Distance, 10^3km	Width, km	Optical Depth	Comments
Galle	1989N3R	41.9	1700	0.0001	High dust content
		49		<0.0001	Outer extent of Galle ring
Le Verrier	1989N2R	53.2	?	0.01	High dust content
Plateau	1989N4R (inner)	53.2		0.0001	Inner edge of plateau
	1989N4R (outer)	59		0.0001	Outer edge of plateau
–		60			Orbit of satellite Galatea
Adams	1989N1R	62.9	15	0.01–0.1	Contains three bright dusty arcs.

The immense difficulty of obtaining images of these rings is understood when we find that exposures of 591 seconds were needed to obtain some of the pictures, and even then the ring material amounts to no more than 12 DN (data number) out of a total range of 256 DN. The rings are composed of very dark material, similar to that of the Uranian rings, with albedoes of a few per cent. To obtain these images at all was a real triumph for the space-craft control and imaging system operation.

It is possible that the narrow rings are confined by the gravitational pulls of the satellites Galatea and Despina, which orbit just inside the Adams and Le Verrier rings, respectively. The satellites would in this case serve to prevent material spiralling in toward Neptune. At the resolution of the Voyager 2 images no shepherding satellites were seen at the outer edge of the ring system, as were seen at Saturn. However, it is quite possible that objects with diameters of less than, say, a dozen kilometres could be present but have escaped detection. At the moment therefore we do not know whether smaller shepherd satellites or some other mechanisms are responsible for controlling Neptune's ring system.

The main rings; 26 August 1989, using the clear filter on Voyager 2's wide-angle camera from 280 000 km. The rings were backlit by the Sun. The bright glare in the centre is due to over-exposure of the crescent of Neptune. Numerous bright stars can be seen in the background.

The three bright arcs contained with the Adams ring are also enigmatical. If this material were allowed to move freely, it ought to spread evenly round the ring in a few years. It is a pity that we cannot watch it over this period of time!

The difficulty of detecting the rings by the radio occultation experiment gives us further evidence that the rings are composed of small particles. Indeed, the particles may actually be smaller than those of the Uranian ring system.

Comparisons with other ring systems

At first sight Neptune, with its retrograde Triton and highly eccentric Nereid, would be expected to have a ring system quite different from those of the other giant planets. In fact, this is not the case. It now seems that the combined systems of rings and inner satellites of Neptune share many of the properties of the systems of Jupiter, Saturn and Uranus.

The Neptunian rings make up a prograde system, lying between the Roche limit and the accretion limit of the planet. (The Roche limit is the distance within which a liquid satellite would break up.) The rings of Neptune make up (1) a narrow and dusty system, quite like the λ ring of Uranus and the F ring of Saturn; (2) diffuse and dusty rings rather like the G ring of Saturn and the Jovian ring; and (3) some confined arcs which are similar to the F and Encke rings of Saturn.

The amount of material in Neptune's rings is 10 000 times less in mass than for Uranus, and many orders of magnitude less than for Saturn. However, the inner satellites of Neptune are substantially larger, and probably more massive, than the equivalent inner satellites of Saturn. At Neptune, the five satellites Larissa, Despina, Galatea, Thalassa and Naiad, whose diameters range between 54 and 190 kilometres, are all within the Roche limit of 77 000 kilometres radius. At Uranus, the nine satellites within the equivalent region range in size from 25–110 kilometres. If we assume that the satellite size distributions would be the same for the different planetary systems, it suggests that there could be many more small undiscovered satellites moving close to Neptune. Moreover, the absence of any moon-shattering events in the recent past (as could have happened at Uranus or Saturn) may also be part cause for the lesser amount of material now in the Neptunian rings.

Both the total masses of the ring systems and their distributions are different among the individual giant planets. If all the material inside the Roche limits for these planets were aggregated into a single body, then for Saturn it would make an object about 195

kilometres in radius, similar to Mimas, which compares with an equivalent 130 kilometres radius body at Neptune and a 75 kilometres body at Uranus.

The relatively large number of very small particles is not unique to the Neptunian rings; it is much the same as for Jupiter's ring and the E ring of Saturn. It is the much larger total amount of dust in Neptune's ring system which presents a problem. Usually these tiny particles do not survive for long, so that if there is so much of it around Neptune there must be an efficient source supply. Moreover, if the material is approximately in equilibrium, then the rates of creation and removal must be similar. It is interesting to note that at Neptune, the material is not confined to the plane of the equator.

Particle impacts were detected at both the ring-plane crossings made during the Voyager fly-by on 25 August, when the space-probe's speed relative to Neptune was more than 76 000 miles per hour. During the inbound pass, the plane was crossed at a distance of about 85 000 kilometres from the centre of the planet; on the outward journey, en route for Triton, the plane was crossed at about 105 000 kilometres. Impacts were recorded 40 minutes before crossing, and reached a peak of 300 per second for 10–15 minutes to either side of the crossing, causing considerable alarm at Mission Control. The distribution of the dust does not seem to be symmetrical with respect to the ring-plane; it spreads both above and below the rings to a distance of more than 50 000 kilometres, with the densest part concentrated within about 700 kilometres. The largest particles are about ten microns in radius. It was a relief when Voyager emerged unscathed from its second crossing!

The origin of the dust is not obvious. At Jupiter, the most likely source of this type of material was thought to be meteoroid bombardment, and this could also apply to the dust at Uranus. Pioneers 10 and 11, which by-passed Jupiter in the early 1970s (and Saturn also, in the case of Pioneer 11) carry dust-detectors, and they have already found evidence of enough interplanetary dust to account for the amounts of fine particles which have been found in the neighbourhood of Uranus and Neptune. But while we can in general account for Neptune's dusty rings, it is not so easy to understand the even larger amounts of dust contained in the Adams and Le Verrier rings; there is much too much to be explained by the meteoroid bombardment theory. Possibly collisions between larger objects could be responsible, and there is a chance that the observed brightening at 57 500 kilometres, near the edge of the Plateau, is one ring feature which may be associated with a belt of small moonlets. It may be significant that something of the same sort has been seen at Uranus.

The Voyager images do not show any finer details

in Neptune's rings, such as the hundred or so rings round Uranus or the detailed internal structure of the D ring of Saturn. This is not to say that it is not present; Voyager could give us only a myopic view. It is possible that the long exposures, together with inevitable image smear, have concealed these very fine details. In fact, there could well be other tiny moonlets and other fine material embedded in Neptune's rings which we have yet to observe.

If this material is steadily created, it must also be steadily removed, and there are several mechanisms for this. At Jupiter, the tiny material is removed mainly by plasma drag; at Uranus, gas drag dominates; at Neptune, whose sparse magnetosphere shows that there is much less plasma and neutral gas, more natural sweep mechanisms are presumably involved.

But how stable are Neptune's rings? Indeed, one of the most exciting issues raised from the Voyager observations is the confinement and stability of the ring arcs. As we have seen, the arcs are brighter sections of the faint but continuous Adams ring. The radial width of the Adams ring is 15 kilometres, and so the differential motion of its inner and outer parts should spread the arc material into a complete circle in about three years. This has simply not happened. The stellar occultation observations made since 1984 prove that the arcs have been stable over a period of at least five years, and they are probably much longer-enduring than that.

One possible explanation is the effect on the nearby rings by the tiny satellite Galatea and its 42:43 outer Lindblad resonance, which could create a 30-kilometre distortion travelling through the arcs. Indeed, this resonance of Galatea coincides with the location of the Adams ring, and its outer 42:43 Lindblad resonance location is only 1.5 kilometres inside this position. This theory suggests that there may be as many as 86 possible stable positions for arcs of azimuthal extents of about 4 degrees, but so far only five arcs have been seen; the leading four span the expected 4 degrees or less, while the trailing fifth arc is 10 degrees long and extends over an unstable region.

An alternative explanation may involve a resonance between Galatea on the Adams ring, and Despina on the Le Verrier ring. This could account for the locations of the inner edges of these rings, recalling the shepherding effect of Cordelia on the inner edges of the ε and possibly the λ rings of Uranus. We must also take into account the possible effects of perturbations caused by changes in the pressure of solar radiation, which would produce effects on a time-scale of around 165 years – the orbital period of Neptune.

Before the Voyager encounter in 1989, Neptune was thought to have only incomplete ring arcs. Now the picture has completely changed. It seems that the rings and the satellite system may not be very old, since a reasonable estimate for the lifetime of the rings, bearing meteoroid bombardment in mind, is a mere hundred million years – and so these relatively young rings have probably been formed from the débris of small satellites in Neptune's neighbourhood.

Satellites of Neptune

Before the Voyager encounter, Neptune was known to have two satellites: Triton and Nereid. Both were strange in their respective ways, since Triton had retrograde motion while Nereid's orbit was remarkably eccentric. Triton is an easy telescopic object; Nereid, with a magnitude of 19, is extremely difficult to see even with a large telescope.

Unfortunately, Nereid was badly positioned when Voyager 2 passed by Neptune; the minimum distance between the satellite and the space-craft was 4 700 000 kilometres, and it is hardly surprising that no good images were obtained. Variations in magnitude, reported by Earth-based observers, were not confirmed during the Voyager fly-by, and from this it seems that either Nereid is nearly spherical in shape or else that it is a slow rotater. Presumably it is icy – the albedo is 0.16 – and very probably its rotation period is not the same as its period of revolution round Neptune, which amounts to over 360 days. The rotation period is sometimes thought to be of the

Nereid: the only picture from Voyager 2.

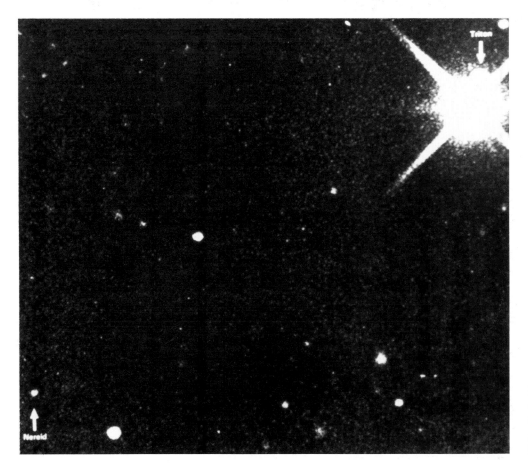

Neptune, Triton and Nereid: G. P. Kuiper, McDonald Observatory. The satellites are arrowed; Triton is very close to Neptune.

SATELLITES

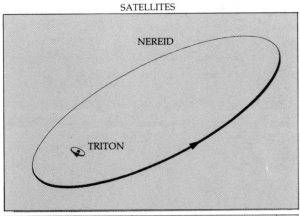

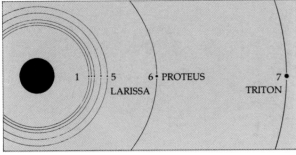

1 NAIAD
2 THALASSA
3 DESPINA
4 GALATEA

order of a month or two, but we do not really know – the rotation could be chaotic. No definite surface features could be identified and, all in all, it is fair to say that our ignorance of Nereid is still fairly complete. Its peculiar, comet-like orbit, which takes it out to well over 9 600 000 kilometres from Neptune, may indicate that it is not a bona-fide satellite, but was acquired by Neptune long after being formed. Note, too, that Nereid is the only satellite to lie beyond the planet's magnetosphere.

To compensate for the failure to tell us much about Nereid, Voyager discovered six new satellites, all close to the planet: Proteus, Larissa, Despina, Galatea, Thalassa and Naiad. (Data are given in Appendix 2.) It is a tribute to the NASA planners that even though the six newcomers were found only as Voyager drew in to Neptune, it was possible to obtain

Orbits of Neptune's satellites.

Newly-discovered Satellites; 24 August 1989. Voyager 2 was then 5 900 000 km from Neptune; the orbital speed of the satellites (over 40 000 km/h) make them show as faint streaks in this 15-second exposure. The satellites have now been named; N3 is Despina, N5 is Thalassa and N6 is Naiad.

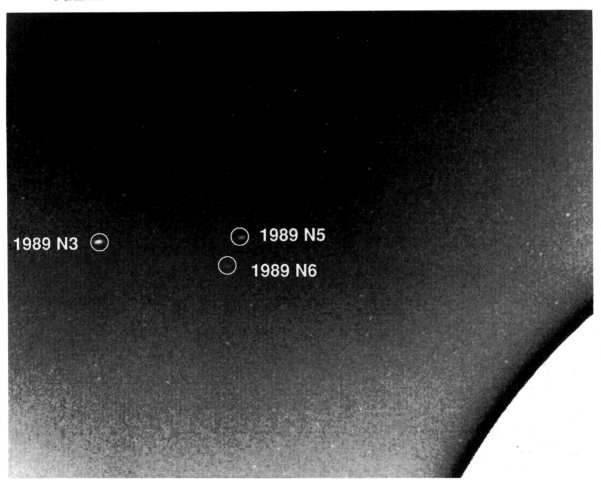

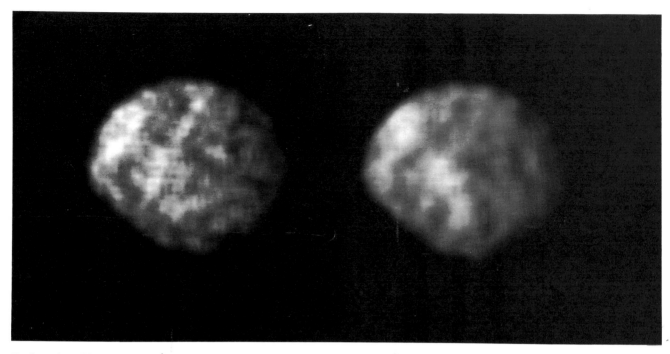

Larissa, from Voyager 2; 26 August 1989. Craters up to 50 km across are shown; the albedo is only about 5 per cent.

Proteus, from Voyager 2.

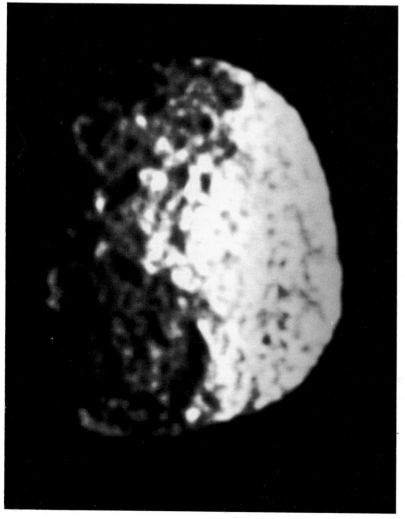

Proteus: Voyager 2 image; 26 August 1989, from 870 000 km. The resolution is 8 km per pixel.

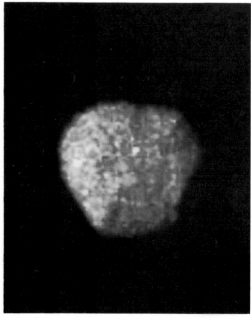

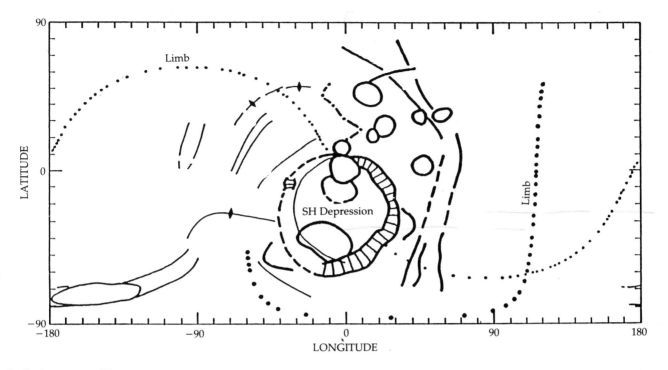

Preliminary map of Proteus.

images of two of them, Proteus and Larissa. Proteus is actually larger than Nereid, but is much more difficult to observe because it is so close to Neptune.

Larissa is somewhat irregular in form, with a rough surface; there seem to be craters from 30–50 kilometres in diameter, but not much else could be made out. Like the other new satellites, it is darkish, and it is reasonable to assume that the rotation period is the same as the orbital period.

Much better views were obtained of Proteus, and it was possible to draw up a map of the surface – or, rather, that part of the surface which was available to Voyager. Its general aspect is dark and bland, so that presumably it has not experienced strong tidal heating; in shape it is decidedly non-spherical, and gives the impression of being somewhat squarish.

The southern part of the Neptune-facing hemisphere is dominated by a large, scarp-bordered depression which has become known as the SHD or Southern Hemisphere Depression. It is 259 kilometres across – and Proteus itself has a diameter of less than twice this value, so that if the SHD were formed by a meteoritic impact it seems that Proteus itself would have been in grave danger of being broken up. The SHD has a circular plan, with raised rims, a flat but rugged floor, and a depth of about 10 kilometres; the width of the inward-facing scarp is 22 kilometres, and the slope has been estimated as about 24 degrees. The depth:diameter ratio is about 0.05,

and it is interesting to compare this with two large craters on Saturn's satellites – Herschel on Mimas, and Odysseus on Tethys, where the depth:diameter ratios are 0.075 and 0.02 respectively. However it was formed, there can be no doubt that the SHD has had a profound influence upon the whole of Proteus' surface. For example there are the linear streaks, depressions from 25–35 kilometres wide, several kilometres deep and up to 100 kilometres long. They form a global network, concentric around one of two axes: the Proteus–Neptune axis, or the axis of symmetry to the SHD. There is also a single bright streak radial to the SHD, which may be either a ridge or else the northern rim of another large, but very degraded, crater.

There are several reasonably well-defined craters. One has a diameter of 87 kilometres, there are a few between 40 and 50 kilometres, and several between 10 and 20 kilometres across. All in all, it seems that Proteus is a primitive, fairly heavily cratered body, more like Mimas in Saturn's system than like Enceladus. We cannot hope for any further information until a new probe passes by Neptune.

Five hours after Voyager 2 had skimmed over Neptune's darkened north pole at 96 000 kilometres per hour came the climax of the whole mission: the approach to Triton. Nobody knew quite what to expect. Methane oceans had been suggested; there was also the possibility that Triton's surface, like that

False-colour image of Triton; 23 August 1989. It is a composite of three images taken through ultra-violet, green and violet filters. The smallest features shown are about 47 km across.

of Titan, would be hidden by cloud. But it soon became clear that many of these speculations had been wrong. First, Triton turned out to be smaller than expected; the diameter is only 2705 kilometres, so that in the 'diameter table' of planetary satellites it comes well below all four of Jupiter's Galileans, Titan in Saturn's system, and our own Moon. If it were smaller than anticipated, it would also have to be more reflective, and therefore colder. The surface

temperature is indeed very low – minus 235 degrees C – so that it is the coldest body ever encountered by a space-craft. The density of the globe also proved to be unexpectedly high, about twice that of water, so that Triton seems to be made up of about two-thirds

Triton; 24 August 1989, from 530 000 km. The resolution is 10 km. The image was made from pictures taken through the green, violet and ultra-violet filters.

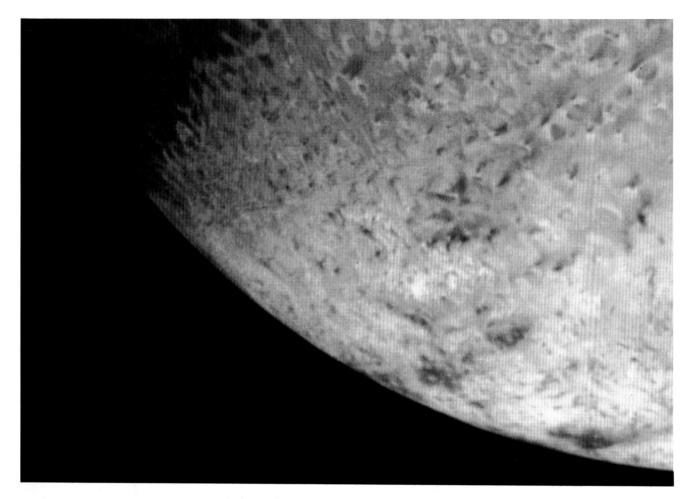

Triton; 25 August 1989, from 210 000 km. The picture is a composite of images taken through the violet, green and clear filters. The largest visible surface features are about 5 km across.

rock and one-third ice. The escape velocity of 1.44 kilometres per second means that an atmosphere can be retained, but it is very tenuous; the surface pressure is about 0.00014 bar (14 microbars), which is a hundred thousand times less than that of the Earth's air at sea-level. Another surprise was the atmospheric composition, which turned out to be 99 per cent nitrogen (N_2), with traces of methane and carbon monoxide. No normal clouds can exist, but there is considerable haze, seen from Voyager above the limb, which extends from 3–6 kilometres above the surface (perhaps slightly more) and is probably composed of microscopic particles of methane or nitrogen ice crystals. Surface winds blow at a rate of around 5 metres per second westward. The temperature in the atmosphere rises to about −173 degrees C at a height of 600 kilometres. This indicates an inversion, but it occurs at a surprisingly high altitude, and the cause of it is not known.

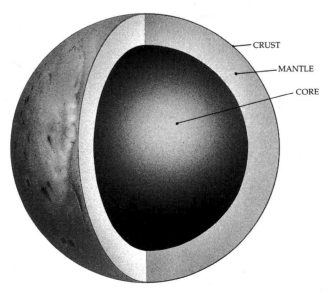

TRITON INTERIOR

Structure of Triton. The interior is probably made up of two-thirds rock and one-third ice; there seems to be a well-defined crust.

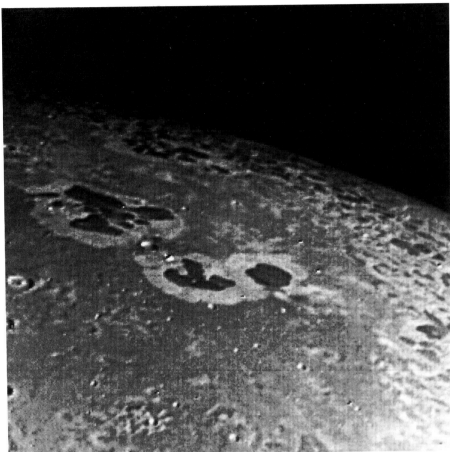

Triton; 24–5 August. The limb cuts obliquely across the middle of the image. The field of view is 1000 km across. Note the strange 'guttæ', whose origin is very uncertain.

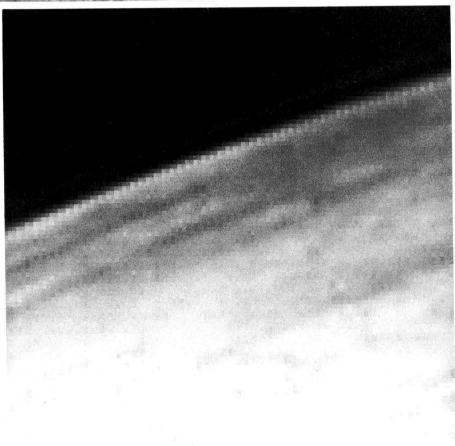

Limb hazes in Triton's atmosphere; 25 August 1989. The main layer begins 3 km above the surface, and is about 3 km thick. A further upward extension of the haze reaches to at least 14 km. The vaguely linear mottling on the surface may be due to shadows from other haze striations. The range was 169 694 km; features are shown to a resolution of 2 km.

The surface of Triton is very varied, and is characterized by a general coating of ice – presumably water ice (H_2O), overlaid by methane and nitrogen ices. Water ice was not detected spectroscopically, but we are sure that it must exist, because methane and nitrogen ices are not rigid enough to maintain surface relief over long periods; they would simply flatten out, whereas water ice is much harder at very low temperatures. Not that there is much surface relief on Triton; there are no mountains or deep gorges, and the total relief cannot amount to much more than a few hundred metres. Neither are there many normal craters. The largest is a mere 27 kilometres in diameter. However, there are extensive flows, some of them at least 80 kilometres wide, which may well be due to ammonia–water fluids.

Fine detail on Triton; the smallest features are about 0.8 km across. The terminator is at the top of the picture, at about latitude 30 degrees N, longitude 330 degrees.

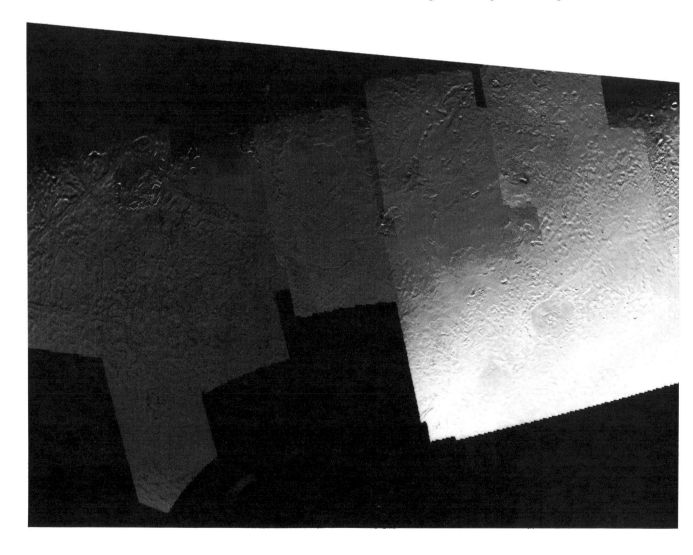

Triton; 27 August 1989, showing two kinds of mid-latitude terrain.

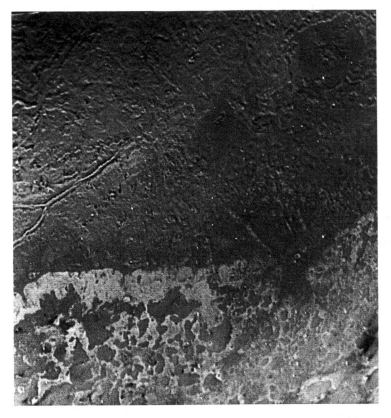

Features on Triton, 25 August 1989. The region is dominated by many roughly circular, polygonal and arcuate features between 30 and 50 km across; the strange intersecting doubled ridge lines are 15–20 km wide and hundreds of kilometres long.

As Voyager 2 approached Triton, it became evident that the colouring was unexpectedly vivid. In particular there was a large pink cap covering the south pole, thought to be due to nitrogen snow and ice. Obviously we have no direct information about the north polar region, which was in darkness at the time of the Voyager fly-by, but at least we have been able to map a considerable portion of the surface. Names for the new features were agreed at the 1991 General Assembly of the International Astronomical Union, and are listed in Table 5.

The Neptune-facing hemisphere is divided into three main regions: Bubembe Regio (western

equatorial), Monad Regio (eastern equatorial) and Uhlanga Regio (polar). Uhlanga is covered by the pink cap, though in some places the underlying geological units show through. The edge of the cap, separating Uhlanga from Monad and Bubembe, is sharp and convoluted. On Triton the seasons are long, and for over a century now Uhlanga has been in sunlight, so that there are signs of evaporation along the borders.

In Uhlanga we find the nitrogen ice geysers which were so unexpected. According to the most plausible theory, there is a layer of liquid nitrogen 20–30 metres below the surface; here the pressure is high enough for the nitrogen to stay in a liquid state. If, however, it starts to rise toward the surface through any weak point in the crust, the pressure will decrease. When it has fallen to about one-tenth that of the Earth's air at sea-level, the nitrogen explodes in a shower of ice and gas, probably about 80 per cent ice and 20 per

The Polar Cap of Triton. Nearly two dozen individual images were combined to produce this view of the Neptune-facing hemisphere. Fine detail is produced by eighteen high-resolution, clear-filter images; colour and contrast have been enhanced to bring out the detail more clearly.

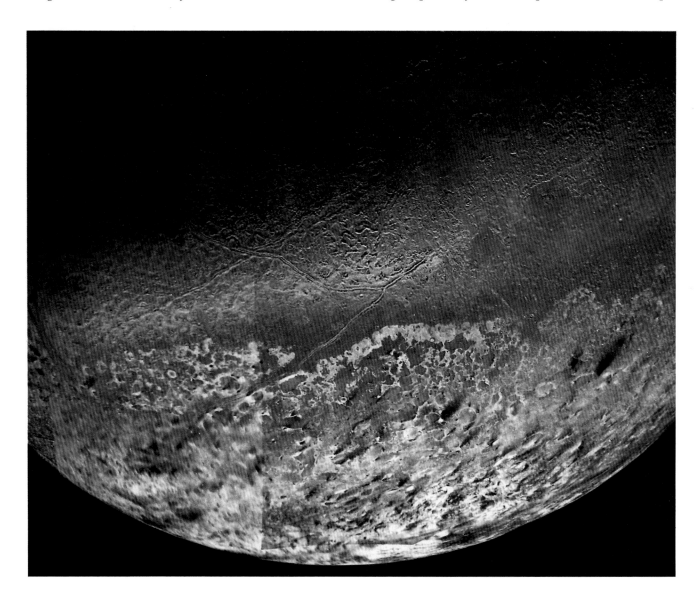

Map of Triton, showing the main named formations.

cent gas. This rushes up through the nozzle of the geyser-like vent at a speed of up to 150 metres per second, and the result is an outburst which sends the material up to a height of many kilometres before it falls back. The outrush sweeps dark material along with it, and this débris is blown downwind, producing streaks of dark material such as Viviane Macula and Namazu Macula, which may be from 15 to 50 kilometres wide and up to 75 kilometres long. It was indeed strange to find activity upon a world as cold as Triton, but the facts seem to be quite clear.

Table 5. *Named features on Triton*

Feature	Latitude	Long. E	Derivation
Abatos Planum	35–8S	35–81	Sacred island in the Nile
Akupara Maculae	24–31S	61–65	Tortoise upholding the world (Indian)
Amarum	26 N	24.5	Quicha water boa (Ecuadoria)
Andvari	20.5 N	34	Fish-shaped dwarf (Norse)
Apep Cavus	20 N	301.5	Dragon of darkness (Egyptian)
Awib Dorsa	1–13 S	73–87	Bushman word for rain (Namibian)
Bheki Cavus	16 N	308	Frog, symbolizing the Sun (Indian)
Bia Sulci	28–48 S	351–14.5	Yoruba, river named for God's son
Boyenne Sulci	18.5–4 S	301–21	Mythological river (Celtic)
Bubembe Regio	25–43 S	285–24.5	Island site of Mukasa temple
Cay	12 S	44	Mayan deity
Cipango Planum	1–26 N	23–46	Legendary island, described by Marco Polo
Dagon Cavus	29 N	354	Fish-shaped fertility god (Babylonian)
Dilolo Patera	26 N	24.5	Sacred lake in Angola
Doro Macula	27–28 S	31–32.5	Mistress of fishing, Sea of Okhotsk
Gandvik Patera	28 N	5.5	Norse sea
Hekt Cavus	26 N	342	Egyptian frog goddess
Hili	56–57 S	24–38	Zulu water sprite
Hirugo Cavus	14.5 N	345	Japanese jellyfish deity
Ho Sulci	0.5–2.5 N	293–317.5	Chinese sacred river
Ilomba	14.5 S	57	Lizi evil watersnake (Zambian)
Jumna Fossae	6.5–20 S	38–50	Hindu river goddess
Kasu Patera	39 N	14	Sacred lake in Persia
Kasyapa Cavus	7.5 N	358	Hindu god Prajapati, as a tortoise
Kibu Patera	10.5 N	43	Mabuiag island of the dead
Kikimora Maculae	28–33.5 S	73.5–82	Slavic swamp-spirit
Kormet Sulci	3 S–33 N	314–345	Norse river of the dead
Kraken Catena	12–15 N	32.5–41	Giant Norse sea-monster
Kulilu Cavus	41 N	4	Babylonian evil fish-man spirit
Kurma	65.5 S	61	Vishnu as a tortoise (Indian)
Leipter Sulci	6–8 N	0–17	Norse sacred river
Leviathan Patera	16–19 N	27–30	Sea-monster upholding the Earth (Greek)
Lo Sulci	3–4.5 N	316–326	Chinese sacred river
Mah Cavus	38 N	6	Persian fish upholding the universe
Mahilani	49–50.5 S	352.5–1	Tonga sea-spirit
Mangwe Cavus	7 S	343	Ila 'The Flooder' (Zambian)
Mazomba	18.5 S	63.5	Chaga, mythical fishes (Tanzanian)
Medamothi Planum	16.5 S–17 N	50–90	Fictional island, 'Nowhere'.
Monad Regio	30 S–45 N	330–90	Chinese symbol of duality
Namazu Macula	24–26 S	12–16	Mythical fish (Japanese)
Ob Sulci	19 S–14 N	325.5–327	Ostiak river entering the Underworld
Ormet Sulci	18 S–29.5 N	328–351	Norse river of the dead
Ravgga	3 S	71.5	Fortune-telling fish–god (Finnish)
Raz Fossae	1–11 N	15–27	Breton bay of souls
Rem Maculae	11.5–15 N	348–351	Fish weeping fertile tears (Egyptian)
Ruach Planitia	24–31 N	19.5–28	French isle of winds
Ryugu Planitia	3–7 S	25–28.5	Japanese undersea dragon palace
Set Catena	21–23 N	35–40	Egyptian evil water monster
Sipapu Planitia	2–6.5 S	32–39	Hole or lake of emergence (Pueblo)
Slidr Sulci	5 S–35 N	312.5–13	Norse river of daggers and spears
Tangaroa	25 S	65.5 E	God of fishing and the sea (New Zealand)
Tano Sulci	23–41 N	327–359.5	Yoruba, river named for the son of God
Tuonela Planitia	36–42 N	7–19.5	Finnish realm across black river
Uhlanga Regio	60–0 S	285–85	Zulu reed, from which men issued
Ukupanio Cavus	35 N	23	Hawaiian shark god
Vimur Sulci	3–12 S	47–77	Norse river Elivagr
Viviane Macula	29, 5–32 S	34.5–38.5	Female companion of Merlin (Welsh)
Vodyanoy	17 S	28.5	Water spirit (Finnish)
Yasu Sulci	2 S–5 N	332.5–0	Heavenly river of peace (Japanese)
Yenisey Fossa	18 S–17 N	52–56	Mythical holy river (Siberian)
Zin Maculae	21–27.5 S	65–71.5	Water spirits (Nigerian).

North of the polar cap there is a darker, redder 'edge'; the colour may be due to the action of ultra-violet light upon the methane. Running across this darker region, more or less parallel to the edge of the cap, is a rather bluish layer; it may be that tiny crystals of methane ice scatter the incoming sunlight and produce a bluish cast – much as our own blue skies are due to the scattering effect of the Earth's atmosphere.

Monad Regio is partly smooth, with hummocky and knobbly terrain. There are rimless or shallow-rimmed pits (pateræ) and graben-like troughs (fossæ), with strange, mushroom-like features such as Zin and Akupara; they have been called 'guttæ',

but their origin is very uncertain, and nothing else quite like them is known elsewhere in the Solar System. Of special interest are the low-walled plains or lakes such as Ruach and Tuonela; they are edged with terraces as though the original level has been changed several times due to repeated melting, ice-vulcanism, and re-freezing. Their interiors are flat and smooth. Most ices would flow like glaciers, but water ice would not, because of the low temperature;

South Polar terrain of Triton; 25 August 1989, showing about 50 dark plumes or 'wind streaks' on the icy surface. The plumes originate at dark spots a few km across, and may be up to 60 km long. It is thought that they must be interpreted as geysers.

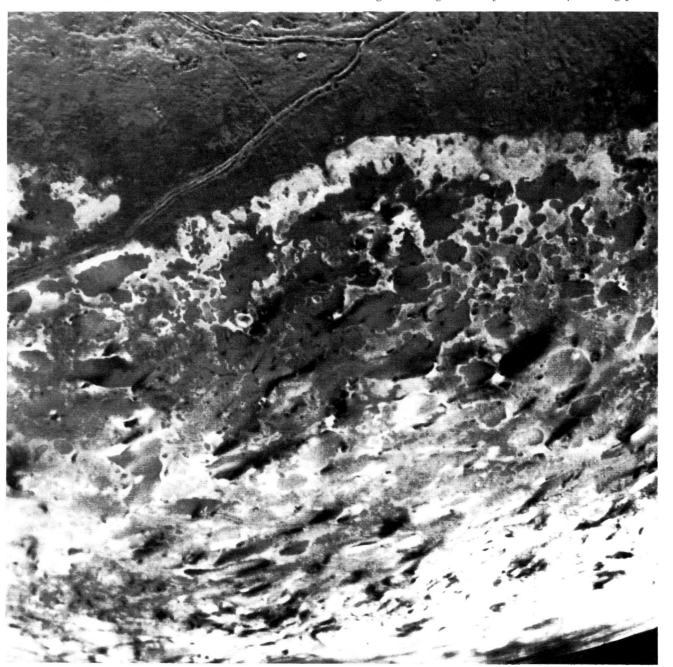

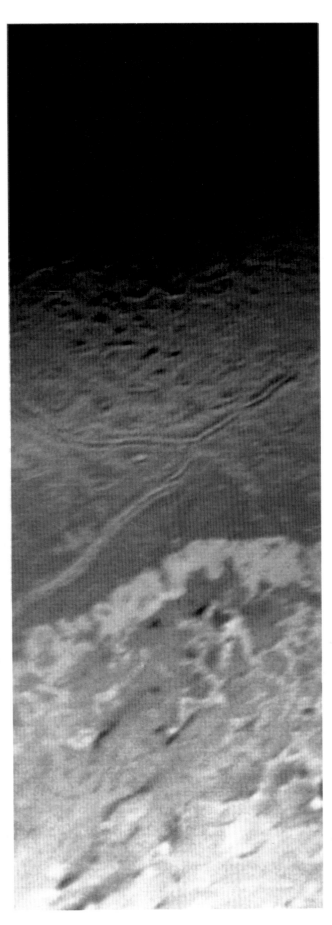

we may therefore assume that water is the main material from which the lakes were formed. Much of the higher part of Monad Regio is covered by undulating, relatively smooth plains.

Bubembe Regio is characterized by the so-called cantaloupe terrain (a name given to it from the superficial resemblance to the skin of a melon). It is crossed by fissures which meet in large X or Y junctions; for example, at the crossing-point of Slidr and Tano Sulci. Liquid material, presumably a mixture of water and ammonia, seems to have forced up through some of the fissures, producing central ridges, and in some cases the material has flowed out on to the adjacent plain before freezing there. The cantaloupe areas are probably the oldest parts of Triton's surface; the pits may be collapse features, while the paterae on the higher smooth regions could be caldera-type vents. The eastern part of Bubembe Regio contains low-

False-colour terrain map of Triton; Voyager 2. Violet, green and ultra-violet images of the satellite were projected into cylindrical coordinates and combined to produce the map.

'Frozen lake' on Triton; 25 August 1989. The picture is about 500 km across, and has a resolution of 900 metres per pixel.

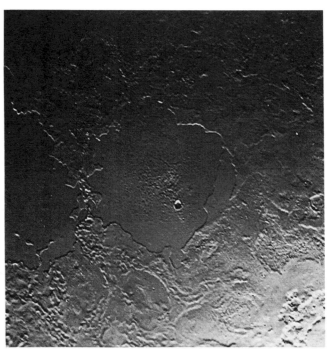

Part of the northern hemisphere of Triton; 27 August 1989, from 80 000 km. The Sun is just above the horizon, so features cast shadows which accentuate surface relief.

Below *A computer-generated perspective rendering of one of the 'frozen lakes' on Triton, as it would appear if viewed from the north east. The image was obtained by geometrically reprojecting part of a Voyager high-resolution frame taken Aug. 24, 1989, when the space craft was about 181 800 km from Triton. Information about surface topography needed to generate the oblique view was obtained from the same image by the technique called photoclinometry (also called 'shape from shading'). In the method, a computer is used to construct a map of the elevation at every point on the image. The computer repeatedly adjusts the map until its appearance (calculated using the known light-scattering properties of Triton) matches the actual image as closely as possible, with bright areas sloping toward the Sun and darker areas sloping away. The topography was vertically exaggerated 20 times in producing this perspective view. Actual relief in the region has a maximum range of about 1 km in the 13 km diameter impact crater visible in the centre of the image. The 'lake' floor, approximately 200 km in diameter, is extremely flat and probably was formed by the volcanic eruption of ice lavas of very low viscosity. The bench visible in the foreground may be a remnant of earlier flooding to a level about 200 m higher than the present 'lake' floor.*

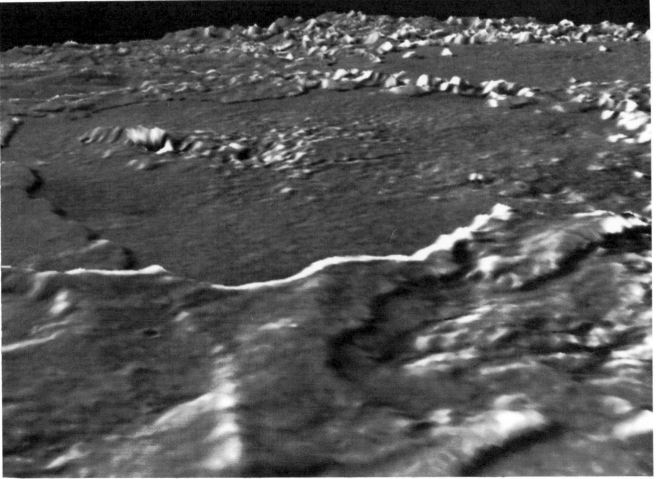

walled circular pits with diameters of around 30 kilometres; some of them have virtually no walls

Triton's unique surface adds force to the argument that it cannot always have been a satellite of Neptune. It may well have begun as an independent body, of the same type as Pluto and Charon (or, for that matter, Chiron) but was then captured by Neptune. The initial orbit would have been eccentric, but over a period of perhaps a thousand million years the path would have been forced into a circular form; there may have been a dense atmosphere, and certainly the interior will have been tidally flexed and heated, so that there was outflowing together with marked surface activity. Obviously we cannot be con-

fident about the situation so long ago, but the capture theory would undoubtedly go a long way toward accounting for some of Triton's odd characteristics.

Quite apart from the nitrogen ice geysers, Triton's surface is probably variable on a larger scale. Southern midsummer will not fall until around the year 2006, and there will presumably be major changes in the cap – and also in the northern polar region, which is at present plunged into its long win-

Linear features on Triton: Voyager 2; 25 August 1989, from 130 000 km. The smallest detail shown is about 2.5 km across. The ridge in the centre of the graben is probably ice which has welled up by plastic flow in the floor of the graben.

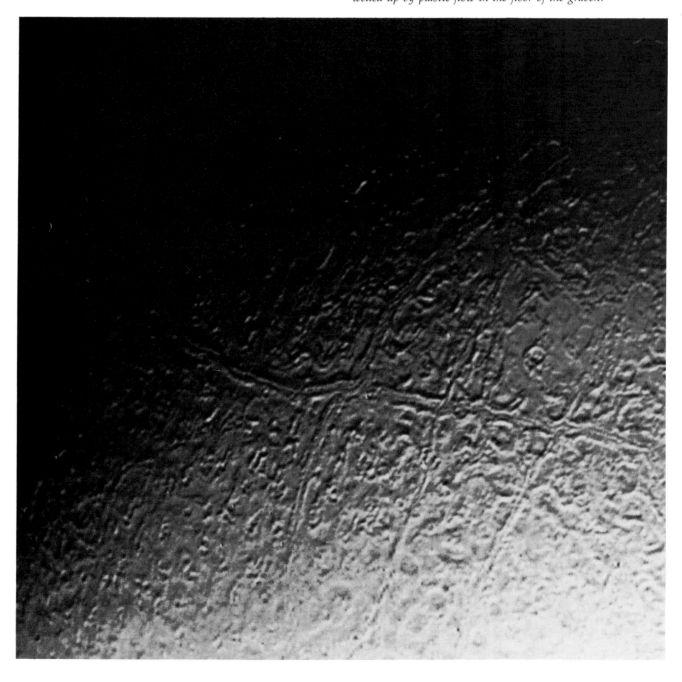

Final view of Triton: Voyager 2; 25 August 1989, from 90 000 km. The phase angle was 155 degrees, so that only a thin crescent of Triton's South Polar Region can be seen. The image was assembled from pictures taken with the green, blue and violet filters of Voyager's wide-angle camera.

ter. It would indeed be fascinating to take another look at this remarkable world in a few years' time, but Voyager 2 can help us no more; by 29 August 1989 it was already 7 000 000 kilometres beyond the Neptunian system, and Triton had become nothing more than a speck in the distance. Before learning more, we must resign ourselves to a long wait.

Beyond Neptune

Until well after the discovery of Neptune, the Solar System was assumed to be complete. Yet there was still something unexplained about the movements of the two outer giants, and as early as 1877 there were suggestions that yet another planet might be involved. At the United States Naval Observatory, David Todd went so far as to carry out a systematic search, but without success. Another investigation was made shortly afterwards by Camille Flammarion, the French astronomer who was not only a skilled popular writer but was also the founder of the Société Astronomique de France. Flammarion's reasoning was based on the movements of comets.

Comets are flimsy things; even a major comet has only a small, icy nucleus, which begins to evaporate as the object draws sunward, producing a coma and (often) a tail or tails. Comets are of two basic types: periodical, and non-periodical. The periodical comets have revolution periods of between 3.3 years (Encke's Comet) and a few centuries; in most cases their orbits are much more eccentric than those of the planets. Dozens of them are known, though only one (Halley's) can ever become a conspicuous naked-eye object.

Flammarion knew that many short-period comets have their aphelia, or greatest points of recession from the Sun, at about the same distance as the orbit of Jupiter. Therefore, he reasoned, Jupiter must have been involved in their formation. There was a similar though much less populus 'family' associated with Saturn. Flammarion calculated that there was another 'family' whose members had aphelion distances well beyond the orbit of Neptune, and this, to him, indicated that there ought to be a planet there. One German astronomer, T. Grigull of Münster, even gave this hypothetical planet a name: Hades. He believed it to be about the same size as Uranus and Neptune, and to have a revolution period of 360 years.

However, no planet came to light, and little more was done until the early part of the twentieth century. The man responsible for the new interest was Percival Lowell, founder of the great observatory at Flagstaff in Arizona. Nowadays Lowell is remembered mainly because of his extravagant theories about Mars; he believed that the Red Planet was the seat of a brilliant civilization, whose engineers had built a vast network of canals to carry water from the icy poles through to the warmer equatorial regions where the 'Martians' lived. Alas, the canals do not exist; but it would be unfair to dismiss Lowell as an eccentric, because we knew much less about the Solar System in his time than we do now.

Lowell was an excellent mathematician, and he took up the problem of the outer planets in very much the same way as Adams and Le Verrier had done so long before. His task was more difficult inasmuch as the proposed planet (Planet X) was bound to be very faint; on the other hand, Lowell could make use of photography – the plan being to photograph the same area of the sky on different nights to see which star-like point had moved. Lowell worked out a position for Planet X, and, unlike Adams or Le Verrier, began to make a personal search, using the powerful 61-centimetre refractor at Flagstaff. Planet X was believed to have a mass seven times that of the Earth, with a period of 282 years and a rather eccentric orbit; the date of the next perihelion passage was given as 1991.

Between 1905 and 1907 various searches were made, but Planet X refused to show itself, and a second search from Flagstaff, carried out in 1914 by C. O. Lampland, was equally fruitless. On Lowell's death, in 1916, the hunt was given up. Yet Planet X was not entirely forgotten; in 1909 another American astronomer, W. H. Pickering, made fresh calculations, and proposed a planet with a mass twice that of the Earth and a period of 373.5 years. On the basis of his results, which were not very different from Lowell's, Milton Humason, at Mount Wilson, undertook a photographic search in 1919, but with no more luck than before.

Then, in 1929, the Director of the Lowell Observatory, V. M. Slipher, decided to try again. He obtained a fine 33-centimetre refractor specially for the task, and called in a young amateur, Clyde Tombaugh, who had come to his attention by sending in some excellent planetary drawings. Tombaugh arrived at Flagstaff, and began work. Success was not long delayed. Early in 1930 he took plates upon which a dim image was found, moving in the expected way. By March it was clear that the new planet had been located, and the announcement was made on the 13th of the month – exactly 149 years after the discovery of Uranus and 78 years after Lowell's birth.[†]

Various names were proposed. One was 'Minerva', and this might well have been adopted but for the fact that it was suggested by T. J. J. See, a former Lowell observer, who was not exactly popular with his colleagues. Finally 'Pluto' was chosen, following a proposal by an Oxford schoolgirl named Venetia Burney (now Mrs Phair). It was appropriate enough; Pluto was the God of the Underworld, and the planet named after him is so far from the Sun that it must be a decidedly gloomy place.

---

† Clyde Tombaugh, now in his eighties, is as active as ever. In 1980, just before the conference held at Las Cruces to commemorate the fiftieth anniversary of the discovery, he invited one of the present authors (P.M) to collaborate with him in writing the official book, and it was a great honour to accept: the book is entitled *Out of the Darkness: the Planet Pluto* (Stackpole Books, USA, and Lutterworth Press, London, 1980).

Clyde Tombaugh. Photograph by Patrick Moore, 1980 – fifty years after Tombaugh discovered Pluto.

One reason why earlier searches had been unsuccessful was that Pluto was much fainter than had been expected. Even when at its brightest the magnitude is only just above 14. Ironically, it was subsequently found that Pluto had been recorded on two of Lowell's plates – and also twice by Humason from Mount Wilson, though in the latter case the planet was missed because of sheer bad luck; one of Pluto's images fell on top of a star, and the other on top of a flaw in the photographic emulsion.

Tombaugh had been carrying out his search in a systematic way, without placing too much reliance on the position given by Lowell, but it was undeniable that Pluto had been found not far from the predicted place, and all seemed to be well. Yet almost at once uneasy doubts started to creep in. For one thing, Pluto had an exceptional orbit, both unusually eccentric and unusually inclined; for part of its 248-year revolution period it is closer-in than Neptune. More importantly, it was found to be small. It seemed to be no larger than the Earth, and successive measurements reduced the diameter still further. The

present value, which is certainly more or less precise, is 2324 kilometres – not only smaller than the Moon, but also smaller than Triton.

Determination of the mass was difficult, but became much easier after 1977, when J. W. Christy at the United States Naval Observatory established the existence of the satellite, Charon, which is a mere 19 000 kilometres from Pluto and is 'locked' with it inasmuch as the revolution period is the same as Pluto's axial rotation period (6 days 9 hours 17 minutes). The ways in which Pluto and Charon moved relative to each other gave the combined mass of the two bodies, which turned out to be less than 19 per cent of that of the Moon.

This was strange. An insignificant system such as this could not possibly have a measurable effect upon the movement of a giant planet such as Uranus or Neptune, so that something was badly wrong. Either Tombaugh's discovery had been fortuitous, or else Pluto was not the planet which had been predicted.

Spectroscopic investigations have told us a little about this curious pair. Pluto has a surface covered at least partly with methane frost, while Charon's weaker gravity has allowed the methane to escape, exposing a layer of water ice. This was established with the help of Nature; during the mid- and late-1980s the orbits of Pluto and Charon were placed so that the two periodically passed behind each other as seen from Earth, and the light from them could be studied separately (a situation which will not recur for many decades). Pluto was also found to have a thin but fairly extensive atmosphere, though Charon lacked anything of the sort. Pluto's atmosphere may be transient; perihelion was passed in 1989, and as the distance from the Sun increases the temperature will fall until the next aphelion (2114) and the present atmosphere may condense on to the surface. If so, then Pluto may be devoid of atmosphere for part of its long 'year'.

All in all, Pluto seems quite inadequate to be classed as a true planet, and as early as 1936 R. A. Lyttleton, at Cambridge, suggested that it might be a former satellite of Neptune which broke away for some reason and moved off in an independent path. This intriguing theory was discarded after the discovery of Charon, but there may still be a link with the Neptunian system, because of a possible association between Pluto and Triton.

As noted, Triton has retrograde motion round Neptune, and this probably indicates that it is not a bona-fide satellite; it was captured well after its formation. If so, then Triton and Pluto may be similar in composition, and may be the largest representatives of other bodies of the same sort in the far reaches of the Solar System. There could even be a link with Chiron, which is admittedly much smaller than

Triton or Pluto, but is easier to locate because it is so much closer-in to the Sun.

It will be some time before we can find out much more about Pluto and Charon, because none of the current space-craft can go anywhere near them, and the dispatch of a special Pluto probe has yet to be planned. The only practicable way is to send the probe past Jupiter, and use the gravity-assist technique. This would be possible for a launch in the year 2002, but whether sufficient funds will be forthcoming seems to be highly dubious.

In any case, it is quite definite that Pluto is not Planet X; so what is the current situation in respect of the hypothetical trans-Neptunian body?

Comets have given us little help. In 1950 K. H. Schütte of Munich returned to Flammarion's old theory of a 'comet family' associated with Planet X, but there was no indication of where Planet X itself might be, and unless there is at least some idea of a position it seems that a search would be as difficult as trying to track down a single new sand-grain in the Sahara. In 1972 K. A. Brady, in America, investigated the movements of Halley's Comet, which has retrograde motion and a period of 76 years, with an aphelion well beyond the orbit of Neptune. Brady's proposed planet was also retrograde, with a mass greater than that of Saturn, a period of 512 years, and a magnitude of above 16. He even gave a position for it, in the constellation Cassiopeia, but searches carried out by both professional and amateur astronomers, including one of the present authors (P.M) proved to be fruitless, and it now seems that Brady's calculations were unsound. The presence of a Saturn-sized retrograde planet would indeed set theorists some serious problems!

Very recently (1992–3) several small bodies have been found moving well beyond the orbit of Neptune. These seem to be around 150 kilometres in diameter, and to be in the nature of planetesimals; probably many more await discovery, and the existence of what has come to be called the Kuiper Cloud is regarded as probable. However, these small objects have no bearing on the problem of Planet X.

If we neglect Halley's Comet, everything depends upon the tiny irregularities in the movements of Uranus and Neptune. But do these irregularities really exist? They are so tiny that they could easily be swamped in errors of observation. There is some evidence that they used to be more marked than they are now, and this has led J. Anderson, of NASA, to make a new suggestion. If Planet X has a long period and a very eccentric orbit, its effects will be noticeable only when it is at or near perihelion; when it is in the far part of its orbit, the perturbations produced on Uranus and Neptune will become too slight to be measured at all. Anderson's Planet X has a revolution period of from 700 to 1000 years, and at the present time it is too remote to make its influence felt.

Obviously there are many possibilities, but we have to admit that as yet we do not know where Planet X is – or even if it exists at all. However, one intriguing 'long shot' involves the probes which are now on their way out of the Solar System: Pioneers 10 and 11, and Voyagers 1 and 2. If all goes well, contact with them will be maintained for some years yet. Suppose that when one of these space-craft has left the orbits of Neptune and Pluto far behind, it encounters Planet X sufficiently closely to have its orbit changed? This could be measured from Earth, and might provide us with the vital clue.

The trouble is that the whole idea depends upon Planet X being in just the right place at just the right time, and the chances of this happening are slender. So for the moment, the puzzle of Planet X remains. If it is a true member of the Sun's family, then sooner or later it may be found – probably later rather than sooner.

Farewell, Voyager 2!

On the day after the official end of the Voyager 2 encounter with Neptune, the present authors presented a television programme in which the results of the mission were summarized. The TV cameras were set up outside the main building of the Jet Propulsion Laboratory. Beside us was the notice-board on which the daily progress of Voyager had been plotted – and on this occasion it carried a significant message: 'Bon Voyage, Voyager'. We were saying 'farewell'.

Not that Voyager 2 had completed its task. It was still fully functional, and the cameras would have been quite capable of imaging any more remote planets – if they existed. Of course, this state of affairs cannot last. As Voyager 2 moves away its signals will fade; the power produced by its tiny nuclear unit will become weaker, and eventually all contact will be lost.

This has not happened yet. It seems that Voyager 2 should remain in touch until about the year 2020, by which time it will have reached the edge of the heliosphere, i.e. that part of the Galaxy in which the Sun's influence is dominant and in which the so-called solar wind, produced by streams of particles sent out by the Sun, is detectable. During the final part of its programme Voyager 2 has been designed to send back data of all kinds, and at the present moment this is precisely what it is doing, but none of these investigations relate to Neptune, which has been left far behind and is quite out of practicable range.

The camera has already been switched off; it can do no more. In 2000 the plan is to shut off power to the scan platform, so as to conserve the last amount of energy in order to keep the remaining instruments working. Long after it has been lost to us it will pass through the Oort Cloud of comets, at a distance of about a light-year; after that, Voyager will enter the final phase of its journey between the stars. Unless it is destroyed by collision with some solid object, or is tracked and captured by an alien civilization, it will continue moving. It may still be wandering in the Galaxy after our own civilization on Earth has died out.

So far as we can tell (and obviously, all calculations of this sort must be very arbitrary), Voyager 2 will not make a close approach to another star for thousands of years. We are confident that we have identified all the nearby stars – that is to say, those within 20 light-years or so – and there is only a slight chance that we have missed a star whose luminosity is so low that it is to all intents and purposes undetectable. The NASA mathematicians have listed the stars which will be by-passed within the next 300 000 years. These are presented in round figures, in Table 6.

All these stars are dim red dwarfs apart from Alpha Centauri, the nearest bright star to the Sun – 4.3 light-years away from us – and of course Sirius, which is

Table 6. *Stars to be by-passed by Voyager 2 during the next 300 000 years*

Year, AD	Star	Distance from Voyager to star, light-years
8 600	Barnard's Star	4.0
20 300	Proxima Centauri	3.2
20 600	Alpha Centauri	3.5
23 200	Lalande 21185	4.7
40 100	Ross 248	1.7
44 500	DM − 36° 13940	5.6
46 300	AC + 79° 3888	2.8
129 000	Ross 154	5.8
129 700	DM + 15° 3364	3.5
296 000	Sirius	4.3

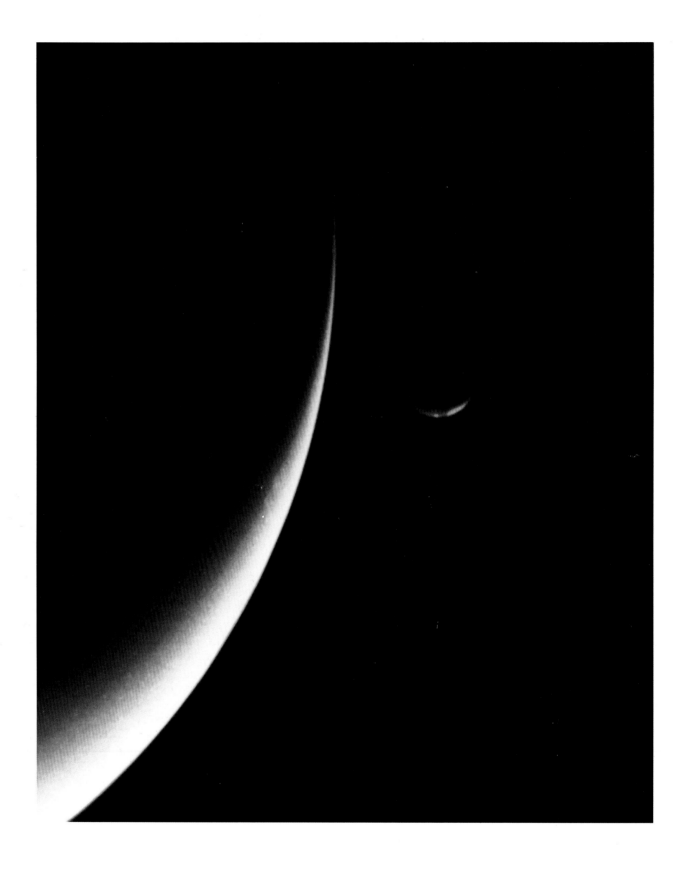

VOYAGER 2
LAUNCHED AUGUST 20 1977
DISTANCES
EARTH 2,758,530,924 MI
NEPTUNE   4,097,124 MI
SPEED TO NEPTUNE 37,582 MPH

POSITIONS   AUG 29

MAGELLAN
LAUNCHED MAY 4 1989
DISTANCES
EARTH   26,629. 5   MI
VENUS   103,814.431   MI
SPEED REL SUN 79,483 MPH

GALILEO LAUNCH — 44 DAYS

'Bon Voyage, Voyager': *The final poster set up at the Jet Propulsion Laboratory, wishing Voyager 2 all good luck on its endless journey.*

Facing page Crescents of Neptune and Triton; 28 August 1989, three days after closest approach to Neptune. Voyager was 4 860 000 km from Neptune and 5 220 000 km from Triton. Colour was produced using images taken through the clear, orange and green filters of the narrow-angle camera.

the equal of 26 Suns, and lies at a distance of 8.6 light-years. If Voyager 2 had flown directly to Sirius, moving at the top legal speed permissible in the United States – 55 miles per hour, or 88 kilometres per hour – the journey would have taken almost 179 000 000 years. Even so, Voyager's minimum distance from Sirius will be as great as that between the Sun and Alpha Centauri, so that in no sense can it be called a 'close encounter'.

After that – who knows? The chance that Voyager may be found by a far-away civilization cannot be ruled out; after all, there are 100 000 million stars in the Galaxy, and there is no reason to doubt that many of them have planet-families of their own. Therefore, Voyager 2 carries what may be termed an interstellar passport. There is a record made on gold-plated copper, containing 116 photographs of Earth and its inhabitants, 90 minutes of music (ranging from a classical symphony to a pop group) and spoken messages, including one from the then Secretary-General of the United Nations, Dr Kurt Waldheim; there are also various sounds, such as the crying of a baby. Obviously, this assumes that our alien friends will know all about record-players!

It would be pleasant to think that Voyager 2 might end its career in a cosmic museum. It has, after all, been probably the most spectacularly successful space-craft we have ever launched. So let us join in the farewell message: *Bon Voyage, Voyager!*

Appendices

Appendix 1. *Data for Neptune*

Parameter	Value
Distance from Sun, 10^6km (a.u.): mean	4496.7 (30.058)
maximum	4537 (30.316)
minimum	4456 (29.800)
Sidereal period, years (days)	164.8 (60 190.3)
Rotation period	16 h 3 m
Mean orbital velocity, km/s	5.43
Axial inclination	28°48′
Orbital inclination	1°45′19″.8
Orbital eccentricity	0.009
Diameter, km: equatorial	50 538
polar	49 600
Apparent diameter from Earth	max. 2″.2, min. 2″.0
Reciprocal mass, Sun = 1	19 300
Density, water = 1	1.77
Mass, Earth = 1	17.2
Volume, Earth = 1	57
Escape velocity, km/s	23.9
Surface gravity, Earth = 1	1.2
Mean surface temperature	−220 °C
Oblateness	0.02
Albedo	0.35
Maximum magnitude	+7.7
Mean diameter of Sun, seen from Neptune	1′ 04″

Appendix 2. *Satellites of Neptune*

Satellite	Mean distance from centre of Neptune, km	Orbital period, days	(hours)	Orbital inclination, degrees	Orbital eccentricity	Diameter, km	Magnitude	Albedo
Naiad	48 200	0.296	(7.1)	4.5	0	54 ± 16	25	0.06
Thalassa	50 000	0.3112	(7.5)	0	0	80 ± 16	24	0.06
Despina	52 500	0.333	(8.0)	0	0	1180 ± 20	23	0.06
Galatea	62 000	0.429	(110.3)	0	0	150 ± 30	23	0.054
Larissa	73 600	0.554	(13.3)	0	0	192	21	0.056
Proteus	117 600	1.121	(26.9)	0	0	416	20	0.050
Triton	354 800	5.877		159.9	0.0002	2705	13.6	0.6–0.8
Nereid	5 514 000	360.16		27.2	0.749	340	18.7	0.16

Triton has a mass of 2.14×10^{22} kg, a density of 2.06 kg/m^3, and an escape velocity of 144 km/s.

Nereid's distance from Neptune ranges between 1 345 500 and 9 688 500 km.

Before the new satellites were named, they were given provisional designations: 1989 N1 (Proteus), N2 (Larissa), N3 (Despina), N4 (Galatea), N5 (Thalassa) and N6 (Naiad).

Triton was discovered by Lassell in 1846, and Nereid by Kuiper in 1949.

Index